Electromanipulation in Hybridoma Technology

A Laboratory Manual

ELECTROMANIPULATION IN HYBRIDOMA TECHNOLOGY

A Laboratory Manual

Edited by

Carl A.K. Borrebaeck and Inger Hagen

M
stockton
press

New York   London   Tokyo   Melbourne   Hong Kong

© Stockton Press, 1989

Published in the United States and Canada by Stockton Press, 1990
15 East 26th Street, New York, N.Y. 10010

Library of Congress Cataloging-in-Publication Data

Electromanipulation in hybridoma technology / edited by Carl
    Borrebaeck, and Inger Hagen.
        p.   cm.
    Includes index.
    ISBN 0-935859-76-4 : $39.95
    1. Hybridomas.   2. Electrofusion.   3. Cell hybridization.
I. Borrebaeck, Carl A. K., 1948-   . II. Hagen, Inger.
    [DNLM: 1. Monoclonal antibodies.   2. Cell Fusion.
3. Electricity.   4. Hybridomas—cytology.   QW 575 E38]
QR185.8.H93E36   1989
616.07′93—dc20
DNLM/DLC
for Library of Congress   89-11317
                                    CIP

Published in the United Kingdom by
MACMILLAN PUBLISHERS LTD, 1990
Distributed by Globe Book Services Ltd,
Brunel Road, Houndmills, Basingstroke,
Hants RG21 2XS, England

British Library Cataloguing in Publication Data

Electromanipulation in hybridoma technology.
    1. Organisms. Hybridomas
    I. Borrebaeck, Carl A.K, 1948-   II. Hagen, I. (Inger)
    575.2′8

    ISBN 0-333-51806-3

U.K. ISBN# 0-333-51806-3
U.S. ISBN# 0-935859-76-4

Printed in the United States of America
9 8 7 6 5 4 3 2 1

CONTRIBUTORS

Mary K. Conrad
National Institute on Drug Abuse
Washington, D.C., U.S.A.

Steven K.H. Foung
Department of Pathology
Stanford University School of Medicine
Stanford, California, U.S.A.

Petra Gessner
Institute of Biotechnology
University of Würzburg
Würzburg, F.R.G.

Mark C. Glassy
Brunswick Biotechnetics
San Diego, California, U.S.A.

Russell Greig
Director, Department of Cell Sciences
Smith Kline and French Laboratories
King of Prussia, Pennsylvania, U.S.A.

Edward Henri
Department of Cell Sciences
Smith Kline and French Laboratories
King of Prussia, Pennsylvania, U.S.A.

Zdenka L. Jonak
Department of Cell Sciences
Smith Kline and French Laboratories
King of Prussia, Pennsylvania, U.S.A.

Mathew M.S. Lo
ICI Pharmaceuticals
ICI Americas, Inc.
Wilmington, Delaware, U.S.A.

Patrick Machy
Department of Cell Sciences
Smith Kline and French Laboratories
King of Prussia, Pennsylvania, U.S.A.

Martin I. Mally
Brunswick Biotechnetics
San Diego, California, U.S.A.

Deborah Matour
Department of Cell Sciences
Smith, Kline and French Laboratories
King of Prussia, Pennsylvania, U.S.A.

Lynette McMillan
Department of Cell Sciences
Smith Kline and French Laboratories
King of Prussia, Pennsylvania, U.S.A.

Susan Perkins
Department of Pathology
Stanford University School of Medicine
Stanford, California, U.S.A.

Steven Trulli
Department of Cell Sciences
Smith Kline and French Laboratories
King of Prussia, Pennsylvania, U.S.A.

Michaela Wander
Institute of Biotechnology
University of Würzburg
Würzburg, F.R.G.

Ulrich Zimmermann
Institute of Biotechnology
University of Würzburg
Würzburg, F.R.G.

PREFACE

Production of mouse monoclonal antibodies is a routine procedure in many laboratories around the world. In most cases, the immortalization of immunocompetent B cells is performed by somatic cell hybridization using the chemical fusogen polyethylene glycol. However, this immortalization procedure is very inefficient with fusion frequencies in the range of $10^{-5} - 10^{-6}$, which is not acceptable, especially when the frequency of immune B cells is low. Today, one of the major obstacles to efficient production of human monoclonal antibodies is the immortalization step.

During the late 1970s pioneering work was reported on the use of electrical pulses to fuse cells. This technique was eventually employed for the production of hybridomas and emerged as an alternative to overcome the shortcomings of existing methods. In particular, one aimed at increasing the low fusion frequencies and later also fusion of antigen-specific B cells, so called directed fusion. Severe technical problems arose immediately, however, when groups outside the pioneers applied this methodology to their individual hybridoma problems, and electrofusion has thus far never played the role of a new general immortalization technology.

The first Workshop on ELECTROFUSION IN HYBRIDOMA TECHNOLOGY was held in Oslo, April 1988 in an attempt to bring the experts and the potential users together to sort out the problems and to answer the question: Is electrofusion a practical alternative in hybridoma formation? The Workshop clearly demonstrated the need for a laboratory manual, a "how-to" description of the recommended procedures, each delineated by an authority in the field. We trust that this manual will be invaluable to investigators who want to evaluate the technology and to apply it to their individual problem in hybridoma formation.

May 1989

Carl A.K. Borrebaeck, D.Sc.
Professor of Immunotechnology
Lund University
Lund, Sweden

Inger Hagen, Ph.D.
Head, Biotechnology Department
Center for Industrial Research
Oslo, Norway

CONTENTS

Preface, vii

1. **Electroinjection and Electrofusion in Hypo-osmolar Solution, 1**
   Ulrich Zimmermann, Petra Gessner, Michaela Wander, and Steven K.H. Foung

2. **Electroporation of Human Lymphoid Cells, 31**
   Zdenka L. Jonak, Patrick Machy, Edward Henri, Lynette McMillan, Steven Trulli, Deborah Matour, and Russell Greig

3. **Formation of Hybridomas Secreting Human Monoclonal Antibodies with Mouse-Human Fusion Partners, 47**
   Susan Perkins, Ulrich Zimmermann, Petra Gessner, and Steven K.H. Foung

4. **Generating Immortalized Immunoglobulin-secreting Human Lymphocytes by Recombinant DNA Technology, 71**
   Martin I. Mally and Mark C. Glassy

5. **B-Cell Hybridoma Production by Avidin-Biotin Mediated Electrofusion, 89**
   Mary K. Conrad and Mathew M.S. Lo

Index, 103

Electromanipulation in Hybridoma Technology

A Laboratory Manual

CHAPTER 1

Electroinjection and Electrofusion in Hypo-osmolar Solution

Ulrich Zimmermann, Petra Gessner, Michaela Wander, and Steven K.H. Foung

I. INTRODUCTION

The discovery of the phenomenon of reversible breakdown of cell membranes in response to electric field pulses of high intensity and short duration paved the way for the development of new tools in bioprocessing and genetic engineering.[1] Electroinjection (electropermeabilization or electroporation) of membrane-impermeable molecules of low and high molecular weight into living cells was the first application of this field pulse technique.[2-4] Research at the Nuclear Research Center, Julich, in the 1970s showed that dyes, proteins, DNA, RNA, and also latex particles could be injected into mammalian cells without deterioration of cellular or membrane functions.[2-9] This method is now used in many laboratories for injection of plasmids into cells and for the formation of transformants.[10-14] The potential of the electric field pulse technique further increased when the first report on fusion of cells at high densities with electrical breakdown pulses was published.[15] This electrofusion method was, however, only fully explored when the required membrane contact between freely suspended cells was achieved by cell alignment in an alternating inhomogeneous field.[16]

The principles and potentials of electroinjection and -fusion for membrane research and for medical and agricultural application have been extensively reviewed under different aspects, and the interested reader is referred to this literature.[17-24] The reader is provided with detailed protocols of very recent modifications in electroinjection and -fusion methods, which have considerably improved and facilitated the application of these pulse techniques to mammalian cells. In these protocols electroinjection and -fusion are performed in *strongly* hypo-osmolar solutions.[25-27] Due to the osmotic stress the membrane permeability is slightly and uniformly increased, presumably because of the dissolution of membrane and cytoskeletal proteins. Obviously, this facilitates electroinjection and -fusion because high and reproducible yields of uptake of membrane-impermeable molecules of low and high molecular weight as well as of hybrids can be obtained.

On the one hand, it is hoped that researchers who are not familiar with physical techniques will develop more confidence in their abilities to recognize and apply this new approach in their research. On the other hand, this chapter will have been useful if it stimulates research into further improvement of the pulse techniques and if pitfalls caused by lack of the required physical background can be avoided. To this end, lists of recently developed equipment, media composition, cell culture, and preparation are described first, followed by electroinjection and -fusion methods used in the formation of murine transformants and hybridoma cells. Guidelines are given in the last section for application of these protocols to other cells, including eggs, bacteria, yeast and plant protoplasts, as well as other mammalian cells.

The electrofusion section of this chapter requires the reading of Chapter 3, in which the conventional electrofusion protocol for the production of hybridoma cells in iso-osmolar solutions is described in detail. Since many experimental steps are identical for both electrofusion protocols, we will discuss only the important modifications.

II. MATERIALS/METHODS

A. Electroinjection

1. Equipment

The Bioject MI (developed by the Institute of Biotechnology, University of Würzburg, manufactured by Biomed, Theres, and distributed by Braun-Melsungen F.R.G under license agreement with the University of Würzburg) has been used to develop the new protocols for electroinjection of foreign molecules (dyes, proteins, DNA, etc.) into living cells under *strongly* hypo-osmolar conditions.

2. Power Supply

The Biojet MI power supply allows the application of exponentially decaying breakdown pulses of very high intensities and of very short durations (Fig. 1-1). The voltages are high enough to electroinject low- and high-molecular weight substances into cells of any size (bacteria, plant, yeast and mammalian cells, eggs) under reversible breakdown conditions. This is an important point that is outlined in "Points to Consider," because lower field strengths or longer pulse duration than about 100 μs lead to irreversible deterioration of most of the cells in the suspension. Simultaneously, suspension volumes of >0.5 ml (with electrode distances of >0.5 cm) can be used for field application. This is a prerequisite for homogeneous field application at normal suspension densities of 10^5 to 10^6 cells/ml. The electronics of the Biojet MI allow both manual as well as programmed operation modes. Twenty electroinjection protocols each with a maximum of 25 program steps can be stored. The facility for automatic operation allows the performance of the entire process of electroinjection in a very short time by presetting all parameters before the experiment.

Features of the Biojet MI

Field strengths: 100 to 15,000 V/cm (corresponding to voltages of 30 kV if a chamber with electrode separation of 0.5 cm is used)

Pulse duration: 5, 10, 15, 20, 40, 1000 μs

Number of pulses: 1 to 10 (manual mode) and 1 to 250 (programmed mode)

Pulse interval: 1 to 59 s with manual mode, 1 min to 60 min with automatic mode

Fig. 1-1. The Biojet MI for electroinjection of membrane-impermeable substances of low or high molecular weights.

Note that in hypo-osmolar electroinjection, only one pulse is required. In iso-osmolar electroinjection, a pulse train has to be applied and therefore the pulse interval must be of the order of 30 s (in order to achieve resealing of the membrane between two consecutive pulses).

Temperature control between 4° and 40°C

All program steps are displayed on a screen.

3. Injection Chamber

The injection chamber is shown in Fig. 1-2. The chamber is reusable. After the stainless steel electrodes are removed, the Plexiglas ring chamber can be sterilized with 70% ethanol. The electrodes can be autoclaved. The sterilized components should be assembled under a laminar flow hood. The cell suspension is injected into the chamber by special (commercially available) pipette tips (see below). Chambers of different volumes are available.

B. Electrofusion

The Biojet CF (developed by the Institute of Biotechnology, University of Würzburg, manufactured by Biomed, Theres, and distributed by Braun-Melsungen F.R.G. under license agreement with the University of Würzburg) has been used for the development of the new electrofusion protocols in strongly hypo-osmolar solutions (Fig. 1-3). The equipment allows the performance of electrofusion in manual or programmed operation modes. The steps of the fusion program are shown in Fig. 1-4. Fifty different electrofusion protocols can be stored. The low output resistance of the device allows electrofusion in weakly conductive solutions (consisting of sugars and small amounts of divalent cations as well as of albumin) and in highly conductive (more physiological) solutions

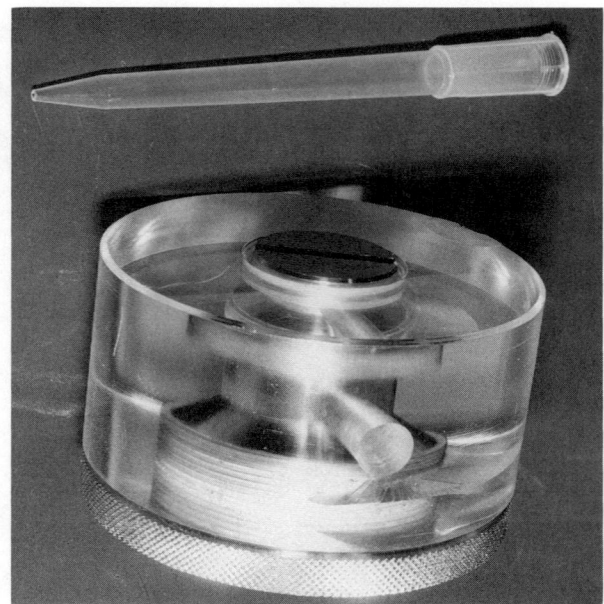

Fig. 1-2. Injection chamber for use in the Biojet MI with slimline long pipette tip.

Fig. 1-3. The Biojet CF for electrofusion of cells.

(containing additional potassium and sodium ions). At present, protocols for electrofusion have only been worked out for weakly conductive fusion media, although the possibility of electrofusion in conductive solutions offers great potential for future applications.

1. Power Supply

A. Cell alignment in an alternating inhomogeneous field (Fig. 1-4): frequency range between 100 kHz and 10 MHz, maximum peak amplitude 15 V (between 100

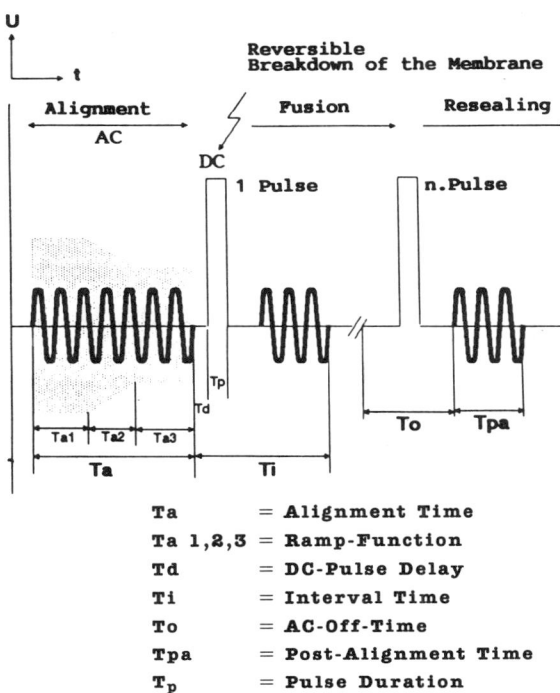

Ta	=	**Alignment Time**
Ta 1,2,3	=	**Ramp-Function**
Td	=	**DC-Pulse Delay**
Ti	=	**Interval Time**
To	=	**AC-Off-Time**
Tpa	=	**Post-Alignment Time**
T_p	=	**Pulse Duration**

Fig. 1-4. Schematic diagram of the sequence of alignment and fusion pulse(s) application.

kHz and 5 MHz), 12 V (between 5 and 8 MHz), and 10 V (above 8 MHz), maximum duration 1000 s; ability to change the amplitude of the alternating field during cell alignment by a ramp function.

B. Fusion (breakdown pulse): strength 1 to 400 V, duration 5 to 100 μs, number of pulses 1 to 9, time interval between consecutive pulses 0.1 to 999 s, AOT (alignment off time) 10 to 999 ms.

C. Postalignment: alternating field as described under item IA. Temperature control between 4°C and 40°C by a ramp function (for heat dissipation during electrofusion in conductive solution or for electrofusion at high environmental temperatures).

D. Various control possibilities are available in order to test proper function of the electrofusion equipment in combination with the fusion chambers (e.g., high-frequency resistance measurements of the connected fusion chamber before and after field application). Resistance measurements before alignment and fusion provide information about correct contact of the fusion chambers with the electronic equipment and about unhindered current flow through the electrodes of the fusion chamber. Resistance measurements after fusion pulse application allow conclusions about successful breakdown of the membranes of the aligned cells because of the conductivity increase of the membrane, due to the release of intracellular ions in response to the fusion pulses.

All program steps are displayed on a screen.

2. Fusion Chambers

Helical chambers, open chambers, and so on are discussed in Chapter 3.

C. Miscellaneous Equipment for Electroinjection and -fusion

Because timing is very critical, it is important to have all equipment available and ready for use.

1. Sterile hood for all cell manipulations.
2. Incubator for cells (37°C with 5% CO_2-enriched atmosphere).
3. Microscope (inverted) for viewing and counting the cells in the culture flasks and in the Neubauer chamber.
4. Horizontal tissue culture flasks (Greiner, Nürtingen, F.R.G., or Nunclon Delta, Denmark; 25 cm² for mouse L cells, 75 cm² for myeloma and G8 cells). NUNC bacterial culture plates 100 × 15 (80 mm² for macrophages, Nunc, Roskilde, Denmark).
5. Particle volume analyzer (Coulter counter) for determination of the cell volume.
6. Centrifuge for centrifugation of the cells at room temperature.
7. Ice bucket for precooling of the injection chamber (only for electroinjection).
8. 1 ml slimline long pipette tips (diameter 0.8 mm); only for electroinjection.
9. Sterile pipettes (10–100 μl, 100–1000 μl, 1000–5000 μl), Pipetmen (Gilson), and so on.
10. Racks for holding the injection and fusion chambers.
11. Trypan blue (0.5%) for viable cell counts.
12. Water bath at 37°C for performance of the resealing process after pulse application.
13. 24 well cloning plates and 96 well microtiter plates (Greiner, Nürtingen, F.R.G., only for electrofusion).

III. MEDIA

A. General Remarks

Although commercially available media can be used for cultivation of cells, special media for electroinjection and -fusion as well as for posttreatment are required. This is because many compounds become toxic when the cell membrane is electropermeabilized. Trace contaminants of heavy metal ions are very critical. Concentrations of the order of nM normally lead to cell death when the cell membrane is in a state of high permeability. Therefore, highly purified ingredients must be used for the media. The best way to remove heavy metal ions is to shake the solutions with Chelex beads (BIO RAD, Richmond, Calif. U.S.A.). These beads consist of a matrix to which chelating agents (e.g., EDTA) are bound. Exposure of the media to these beads removes most of the multivalent ions, including calcium and magnesium ions, which have to be added to the media afterward (see below).

It is important to note that only immobilized chelating agents should be used for purification. Otherwise, uptake of chelating agents (adsorbed to the outer surface of the membrane of the cells) will occur during field pulse application and during the subsequent resealing process. Incorporation of molecules of chelating agents also leads to cell death. This is one reason why electroinjection of DNA molecules frequently fails. Most of the plasmid preparations contain EDTA to inhibit DNAses. Therefore, EDTA has to be removed carefully prior to electroinjection. The best procedure, however, is to perform the preparation of plasmids in the absence of EDTA. The plasmids can be stored at $-20°C$ without destruction.

For the same reasons, wash- and post-pulse media that do not contain phenol red (or other dyes used as pH-indicators) must be used. These dyes are very toxic for cells with permeabilized membranes. Complete growth media with pH indicators should be used only for cultivation of transformants or hybrids after 1 day at the earliest. This time is sufficient to heal any perturbation in the membrane introduced by the breakdown pulse.

B. Electroinjection

Complete Growth Medium

RPMI 1640 medium (Biochrom, Berlin, F.R.G.) supplemented with fetal calf serum (FCS, Boehringer Mannheim, F.R.G.), 5% for mouse L-cells, and 10% for macrophages J774

2 mM L-glutamine (Biochrom, Berlin, F.R.G.)

1 mM sodium pyruvate (Biochrom, Berlin, F.R.G.)

Nonessential amino acids (Biochrom, Berlin, F.R.G.)

Penicillin and streptomycin antibiotics (100 U/ml and 100 μg/ml, respectively; Biochrom, Berlin, F.R.G.)

Media for Cell Passage

For mouse L cells: 0.25% trypsin in phosphate buffer solution (Biochrom, Berlin, F.R.G.)

For macrophages: EDTA-phosphate buffer solution consisting of 137 mM NaCl, 2.7 mM KCl, 8.1 mM $Na_2HPO_4 \times 2 H_2O$, 1.5 mM KH_2PO_4, 5.4 mM tritiplex III, and 5.4 mM EDTA (all compounds analytical grade, Merck, Darmstadt, F.R.G.)

The dry ingredients are dissolved in double distilled water, the pH is adjusted to 7.2 with 5 mM KOH. Then the solution is autoclaved at 121°C for 30 min.

Media for Enzymatic Treatment

Complete growth medium supplemented with 0.1 mg/ml dispase (6 U/mg, grade 1, Boehringer Mannheim, F.R.G.) for mouse L cells

1 mg/ml pronase (Boehringer, Mannheim, F.R.G.) for macrophages J774.A1

Wash Medium

1. Complete growth medium without phenol red (RPMI 1640 without phenol red is supplied by Biochrom, Berlin, F.R.G.)
2. RPMI 1640 without phenol red

Pulse Medium

30 mM KCl (Merck, Darmstadt, F.R.G.)

0.80 mM K_2HPO_4 and 0.35 mM KH_2PO_4 (Merck, Darmstadt, F.R.G.)

Appropriate amounts of inositol (Sigma, Deisenhofen, F.R.G) to adjust the osmolarity of the solution to 75 mOsmol (or 100 mOsmol). Osmolarity should be checked by means of an osmometer (Osmomat 030, Gonotec, Berlin, F.R.G.).

The dry ingredients are dissolved in double distilled water, the pH is adjusted to 7.2 and the solution is autoclaved at 121°C for 30 min.

Appropriate amounts of the foreign molecules have to be injected. In the case of DNA injection, concentrations of 1 to 5 μg DNA/ml are normally required.

Resealing Medium

RPMI 1640 without phenol red (Biochrom, Berlin, F.R.G)

Post-pulse Medium

Mouse L cells: RPMI 1640 without phenol red and RPMI 1640 without phenol red, but with double concentrations of the ingredients of the growth medium, ratio 1:5

Macrophages: Complete growth medium, except with 20% FCS

Selection Medium

Mouse L cells: complete growth medium supplemented with 500 μg/ml G418 (Gibco, Karlsruhe, F.R.G.)

Macrophages: complete growth medium supplemented with 250 μg/ml G418 (until the 5th day) and with 500 μg/ml G418 (after the 5th day)

C. Electrofusion

Complete Growth Medium

RPMI 1640 (Biochrom, Berlin, F.R.G.)

10% FCS (Boehringer, Mannheim, F.R.G.)

2 mM L-glutamine (Biochrom, Berlin, F.R.G.)

2 mM sodium pyruvate (Biochrom, Berlin, F.R.G.)

2 g/l $NaHCO_3$ (Biochrom, Berlin, F.R.G.)

Nonessential amino acids (Biochrom, Berlin, F.R.G.)

50 μM 2-mercaptoethanol

100 U/ml/100 μg/ml penicillin/streptomycin (Biochrom, Berlin, F.R.G.)

Fusion Medium

> 0.1 mM calcium acetate
>
> 0.5 mM magnesium acetate
>
> 1 to 2 mg/ml bovine serum albumin (BSA, Serva 11930) (Note that different charges of albumin can have different effects on the fusion and hybrid yield.)
>
> Appropriate amounts of sorbitol (Merck, Darmstadt, F.R.G.) to adjust the osmolarity of the medium to 75 mOsmol. The osmolarity of the medium has to be determined by an osmometer (Osmomat 030, Gonotec, Berlin, F.R.G.).

Post-fusion Medium

> Complete growth medium without phenol red

Selection Medium

> Complete growth medium supplemented with HAT (Boehringer, Mannheim, F.R.G.; final concentration: 6.8 mg/ml hypoxanthine, 0.088 mg/ml aminopterin, and 1.938 mg/ml thymidine)

IV. CELLS FOR INJECTION AND FUSION

All cells are cultured at 37°C with 5% CO_2. For electric field treatment, cells should be in log phase growth with optimal viability. Dead cells diminish yields in electroinjection and -fusion.

A. Cells for Injection

1. Mouse L Cells

Cells are grown to confluence on the surface of horizontal flasks in complete growth medium. Once confluence has been reached the complete growth medium is decanted. The adhered cells are rinsed with 1 ml of the medium for cell passage for dissolution of the cells from the surface of the culture flask. Then the cells are suspended in 10 ml complete growth medium; 1 ml of this cell suspension is diluted with complete growth medium in a 1:10 ratio and cultured for 1 day. Under these conditions, the cells are in the log phase as required for electroinjection. Just before electroinjection, the complete growth medium is decanted and replaced by complete growth medium containing dispase. After 1 h enzymatic treatment at 37°C and 5% CO_2-enriched atmosphere, the cells are centrifuged for 10 min at 150 *g* at room temperature. The pellet is resuspended and washed with wash medium 1, centrifuged, and washed with wash medium 2. Afterward the cells are transferred into the hypo-osmolar pulse medium.

2. Macrophages

The macrophages (strain J 774 A.1) are grown in bacterial culture plates using the complete growth medium. One day before electroinjection, the culture medium is decanted and the attaching cells are rinsed with the phosphate buffer solution containing EDTA. After 1 min, the cells are removed from the surface of the culture plates by hitting the plates a few times. The cells are then resuspended in 10 ml complete growth medium and split into 4 bacterial culture plates, each of which is filled with 7.5 ml of complete growth medium. After 1 day, the complete growth medium is decanted and replaced by complete growth medium containing protease. After 30 min enzymatic treatment at 37°C, the cells are centrifuged for 10 min at 150 g and washed first with complete growth medium and then with wash medium 2. After the last washing procedure, the pellet is resuspended in the pulse medium.

B. Cells for Electrofusion

1. Mouse Myeloma Cells

The nonsecreting myeloma cell line, SP2/0-Ag 14, is grown in 10 ml complete growth medium at 37°C in an atmosphere supplemented with 5% CO_2. Every 2 to 3 days, the cell suspension is diluted 1:10 and transferred in new flasks for cultivation in order to keep the cells in the log phase for electrofusion. The HAT-sensitivity of selected cells is verified in experiments performed at regular intervals. One day prior to fusion, the cells in log phase are split 1:10 and grown to a density of 3-5 $\times$ 10^5/ml in complete growth medium. Then the cells are harvested by 10 min centrifugation at 150 g (at room temperature).

2. G8 Hybridoma Cells

The murine hybridoma cell line, G8, is prepared as follows[26]: splenic B cells derived from a C57B1/6 mouse are fused to the nonsecreting myeloma cell line SP2/0-Ag 14, using polyethylene-glycol (PEG). The resulting hybridomas are selected in complete growth medium supplemented with HAT. A thymidine-kinase deficient variant of one of these cells is grown for 1 week followed by subcloning in growth medium supplemented with 20 μg/ml of bromo-deoxyuridine (BdUR). HAT-sensitive clones are selected and passaged in BdUR-supplemented growth medium. Control experiments to verify continued HAT sensitivity are performed at regular intervals. For electrofusion, G8 cells are grown and harvested under the same conditions as myeloma cells. Note that the G8 cells adhere to the surface of the culture flask. The adhered cells are removed by rinsing with complete growth medium and simultaneously shaking the culture flask.

3. In Vivo Activated B Cells

Mice (Balbl/C-AnNCr1BR) are immunized with dinitrophenol (DNP) conjugated to *Limulus polyphemus* hemocyanin (Hy). The mice receive two intraperitoneal (i.p.) injections of 100 μg of DNP-Hy, the first in complete Freund's adjuvant and the second in normal phosphate buffered saline (PBS) in a volume of 0.2 ml. At least 4 weeks after the second i.p. injection, the mice are boosted with 100 μg of DNP-Hy in saline intravenously (i.v.). Three days after the i.v. boost, the mice are sacrificed and their spleens are removed aseptically. The spleens are crushed and single-cell suspensions are made. The cells are washed once in complete growth medium, resuspended in 10 ml of 0.8% ammonium chloride, and incubated for 5 min at 37°C in an atmosphere enriched with 5% CO_2 to lyse the red blood cells. Then this suspension is centrifuged for 10 min at 150 g, and the cells are washed in complete growth medium and counted. Splenic T cells are then lysed by resuspension of the cells in 100 μl of 1:200 anti-murine Thy 1.2 in PBS at 37°C and 5% CO_2-enriched atmosphere for 30 min. Afterward, the cells are washed in complete growth media and counted and resuspended in 500 μl of 1:10 diluted guinea pig complement (Flow, McLean, VA) in complete growth medium at 37°C for 45 min in an atmosphere enriched with 5% CO_2 to lyse the T-lymphocytes. Then the cells are washed three times in complete growth medium and counted.

METHOD I: Electroinjection of Plasmids in Hypo-osmolar Solutions

This procedure has been used successfully for the reproducible high yields of transformants of mammalian cells.[27] In these studies, the plasmid DNA pSV-2 neo (conferring resistance to the antibiotic G-418) was electrically injected (for isolation of this plasmid from *Escherichia coli*, see Southern and Berg[29]). Prior to use, the plasmid must be linearized with EcoR1 because the yield of transformants is less—at least in mammalian cells—if circular plasmids are used.[30] High, efficient uptake of plasmids (and also of other foreign, membrane-impermeable molecules) is achieved when the breakdown pulse is applied at 4°C (see below). Most of the mammalian cells cannot be kept for long at this temperature without cell deterioration, particularly if they are suspended in the nonphysiological, hypo-osmolar pulse medium.[27,31] Therefore, time is critical once the cells are transferred into the pulse medium. This implies that the Plexiglas ring chamber must be precooled on ice after sterilization and that the field parameters are preset. The guideline for the selection of optimum field parameters is described in "Points to Consider." For transfection of mouse L cells in hypo-osmolar (75 mOsmol) solutions, a field strength of the breakdown pulse of 4 to 6 kV/cm is optimum. For macrophages, a field strength of 8 kV/cm is required.[27] The pulse duration must be adjusted to 5 μs. Application of a single pulse is sufficient. This contrasts with electrotransfection in iso-osmolar solutions, where at least 3 to 5 pulses are necessary to achieve reasonable yields of transformants.[31]

Procedure

1. Sterilize the transfection chambers by rinsing with 70% ethanol. After evaporation of the alcohol, cool them on ice.

2. In the meantime: prewarm the washing and resealing media to 37°C and pipette 3 ml L-cell resealing medium into centrifuge tubes. Place these tubes into a water bath at 37°C. Also pipette 6 ml of the L-cell post-pulse medium (or 20 ml macrophage-post-pulse medium) into the tissue culture flasks and place them in the incubator. Precool the pulse medium on ice.

3. Add 10 ml of the medium for enzymatic treatment to the adhered cells after decanting the supernatant. Use the dispase medium for mouse L cells and the pronase medium for macrophages. Perform the enzymatic digestion at 37°C in the incubator for 60 min in the case of mouse L cells and for 30 min in the case of macrophages in a 5% CO_2-enriched atmosphere.

4. Preset the required pulse parameters by programming the Biojet MI and precool the electrodes of the equipment to 4°C. For adjustment of the optimum field strength, determine the volume of an aliquot of cells suspended in hypo-osmolar solution by using a microscope or a Coulter counter (in order to calculate the field strength for injection of DNA dependent on the radius of the cells, see "Points to Consider").

5. Harvest the cells after enzymatic treatment by gently rinsing the bottom of the culture flask. Centrifuge the cells in tubes, aspirate the medium from the spun cells, and resuspend the cells in the first washing medium. Repeat the washing procedure with the second, prewarmed washing medium and count the cells in a Neubauer chamber or in a Coulter counter. Resuspend the cells after counting in the precooled pulse medium and adjust the final cell density to 2×10^5 and to 2.5×10^6 cells/ml for mouse L cells and macrophages, respectively.

6. Add 1 μg/ml DNA (in the case of macrophages 5 μg/ml) and mix the suspension by gently shaking the tube.

7. Fill the cell suspension into the ring chamber of 0.5 to 5 ml volume (depending on the amount of cells and of plasmid supply). In this protocol, chambers

with a volume of 1 ml are used. Note that with smaller chamber volumes, the temperature may rise during pulsing. In this case, the temperature reading of the Biojet MI does not reflect the actual temperature in the cell suspension.

Use a 1 ml slimline long pipette tip for filling the chamber, in order to avoid shearing forces on the cells. Fill the chamber slowly from the bottom to the upper edge of the electrode areas in order to eliminate the formation of gas bubbles. (Pulsing in the presence of gas bubbles between the two electrodes can cause the chamber to explode, especially if high field intensities are applied.) Lock the chamber and place it with the inlet upward into the electrode holder of the Biojet MI. The large electrode area must face the cathode thermostat. Slight tightening of the anode (the electrode with the small area) will fix the chamber in position.

8. Start the automatic program for pulsing. Note that if the chamber has been inserted incorrectly, no voltage will be generated. An additional safety circuit ensures that cells can be pulsed only if the front lid is closed and the chamber is in the correct position.

9. After breakdown (electropermeabilization), the cells must be kept at 4°C for at least 2 min before the chamber is removed from the holder. This time interval is required for sufficient uptake of DNA (or other foreign molecules) into the cells by passive diffusion. Longer time intervals result in a decrease of the total number of transformants (Fig. 1-5). Then the chamber is opened and the cells are transferred very carefully into a centrifuge tube containing 3 ml prewarmed resealing medium. These tubes are kept at 37°C in a water bath for about 30 min.

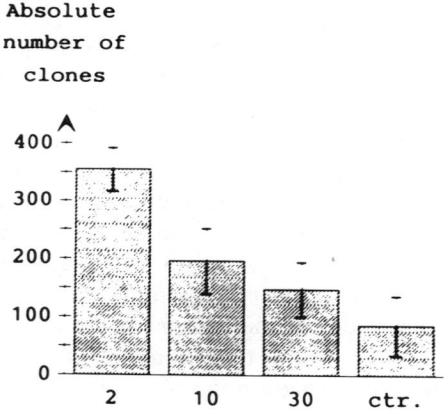

Post-pulse time   [min]

Fig. 1-5. The absolute number of stable transformants obtained after electroinjection of mouse L cells in 75 mOsmol solutions (containing 1 µg/ml pSV-2 neo plasmid) as a function of the post-pulse time at 4°C. Conditions: cell suspension density 2×10^5 cells/ml, single field pulse of 5 kV/cm strength and 5 µs duration. The post-pulse time was varied between 2 and 30 min. The control cells (ctr.) were suspended in iso-osmolar solutions (300 mOsmol) and subjected to a field pulse of 10 kV/cm; post-pulse time 10 min. The data and error bars represent the mean of six (in the case of the controls of four) experiments. (Redrawn from Daümler and Zimmermann.[27])

10. Take aliquots of about 90 μl and mix them with 10 μl trypan blue (0.5%) in order to determine the number of (viable) cells in a Neubauer chamber. Afterward, transfer the pulsed cell suspension into the culture flask containing 6 ml prewarmed post-pulse medium for mouse L cells (in the case of macrophages 20 ml macrophage-post-pulse medium).

11. After 2 days of cultivation, the post-pulse medium is replaced by the appropriate selection medium.

12. Count antibiotic-resistant clones after 8 to 10 days. In the case of macrophages, 10 to 12 days are normally required.

Figures 1-6 and 1-7 show the transformant yield for mouse L cells and macrophages, respectively, compared with iso-osmolar electrofusion conditions.

METHOD II: *Electrofusion of Mammalian Cells in Weakly Conductive, Hypo-osmolar Solutions*

In contrast to iso-osmolar electrofusion,[32] this protocol for hypo-osmolar electrofusion results in very high and reproducible yields of hybridomas.[25,26] Electrofusion must be performed at room temperature (about 25°C). Time is also more critical than with electroinjection because the cells are suspended in solutions containing only very small amounts of divalent cations and no potassium or sodium ions. In addition, the low osmolarity limits cell survival to about 1 to 2 h. Therefore, the parameters for the alignment field and for the field pulse(s) must be preset. Guidelines for the selection

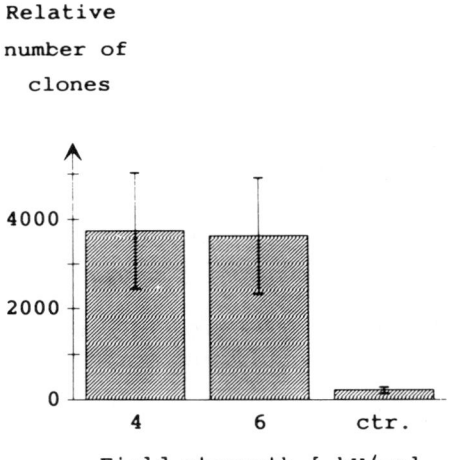

Fig. 1-6. Relative number of stable clones obtained after electro-transfection of mouse L cells in 75 mOsmol solution with field pulses of 4 and 6 kV/cm strength, respectively. (Experimental conditions as in Fig. 1-5.) The numbers of clones are referred to 10^6 cells/ml. The post-pulse time at 4°C was 10 min. The data are the mean of three independent measurements; the bars represent the standard deviations. The control cells (ctr.) were suspended in 300 mOsmol pulse media and treated in the same manner. The field strength was adjusted to 10 kV/cm because of the smaller cell radius (see text for explanation). (Redrawn from Daümler and Zimmermann.[27])

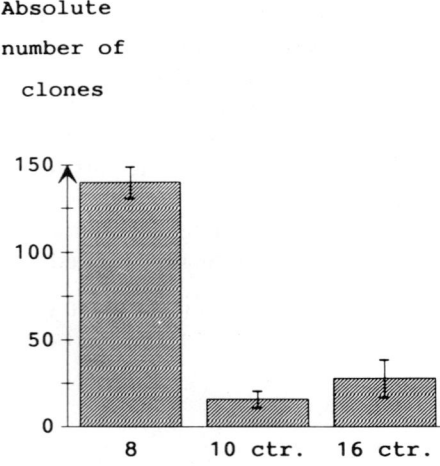

Fig. 1-7. Absolute number of stable clones counted after electro-transfection of J774.A1 macrophages suspended in 75 mOsmol solutions. Conditions: suspension density 2.5×10^6 cells/ml, pSV-2 neo plasmid concentration 5 μg/ml, field pulse of strength 8 kV/cm and 5 μs duration, post-pulse time 2 min at 4°C. The data are the mean of five experiments; the bars represent the standard deviations. The control cells (ctr.) were suspended in iso-osmolar solutions (300 mOsmol) and treated in the same manner, except that the field pulse strength was adjusted to 10 or 16 kV/cm because of the smaller cell volume (see text for explanation). (Redrawn from Daümler and Zimmermann.[27])

of optimum field parameters are given in "Points to Consider." For hypo-osmolar electrofusion, the following field conditions have been found to be optimal (Fig. 1-4):

1. Constant alignment field of 250–300 V/cm amplitude, 1.5. to 2 MHz frequency and 30 s duration.
2. Three consecutive square pulses at 1 s intervals of 1.5 kV/cm strength (in the case of G8 × SP2/0 1.25 to 1.75 kV/cm) and of 15 μs duration.
3. The alignment off time (AOT) of the alternating field during each pulse is adjusted to 10 ms.
4. After the final pulse a constant alternating field of the same amplitude and frequency is applied for further 30 s to keep the fusing cells in close contact.

Recently, it has been shown that variation in the amplitude of the alternating field during the alignment phase can have great advantages over application of a constant amplitude. Preliminary results show that application of an alternating field of an amplitude of 500 V/cm for 4 s followed by an alternating field of an amplitude of 250 V/cm for 20 s leads to preferential alternating alignment of. Lowering of the high amplitude to the low one is electronically achieved by a ramp of 3 s duration (Fig. 1-4).

Procedure

1. Harvest the activated lymphocytes (or G8 cells) and the myeloma cells, wash them with 10 ml complete growth medium, resuspend them in the same volume of complete growth medium, and count the cells in a Neubauer chamber or in a Coulter counter and mix them. For hypo-osmolar electrofusion of G8 cells

with SP2/0 cells, a 1:1 ratio is optimum; whereas for fusion of lymphocytes with myeloma cells, a ratio of 1:2 yields highest fusion and hybrid frequencies. Wash the cell mixture twice in fusion medium (75 mOsmol) and resuspend 6 $\times$ 10^5 lymphocyte/myeloma cells in 200 μl (or 2 $\times$ 10^5 G8/myeloma cells in 340 μl) of the hypo-osmolar fusion medium. Pipette this volume carefully into the (sterilized) receptacle of the precleaned helical chamber. Use a pipette with a large tip diameter in order to avoid the occurrence of shearing forces on the cells. Don't touch the inner surface of the receptacle with the pipette during the filling process.

Slowly insert the helical assembly with electrode spacing of 200 μm with a twisting motion. Once the helix has touched the cell suspension, be sure to keep moving it downward while twisting to prevent air bubbles from forming. Once the electrode assembly is firmly seated in the receptacle, place the chamber in the holder. Plug the two connector cables into the power supply of the Biojet CF and measure the high-frequency resistance of the assembly. Be sure that there is unhindered current flow through the electrodes of the helical chambers (see description of the apparatus above). Note that failure of electrofusion often results from wrong connections of the cables of the fusion chamber to the power supply, or from breaks in the electrode wires (which may occur after many uses), or from insulating layers on the electrode surfaces due to cell debris (cleaning of the electrode assembly is very important).

2. Push the buttons for the automatic program of the preset fusion process. After the fields are shut off, measure the high-frequency resistance of the chamber. A slight increase in conductivity is an indication that the cells were exposed to the field pulse(s). Then gently disconnect the electrodes and keep the chamber at 25°C for 10 min without disturbance. During this time interval, fill the wells of the cloning plates with post-fusion medium and add the feeder layer.

3. Open the chamber gently and rinse the helical assembly and the receptacle with 1 ml post-fusion medium. Fill each of the 4 wells of a 24-well cloning plate with 1 ml post-fusion medium and add 0.3 ml of the diluted pulsed cell suspension into each well. In the case of lymphocyte/myeloma fusion each well contains 1 to 2 $\times$ 10^4 peritoneal macrophages as feeder layers. After 24 h at 37°C and 5% CO_2-enriched atmosphere, replace the post-fusion medium by 1 ml selection medium. This must be done very carefully.

4. Count the clones of viable hybrids 5 to 7 days later under an inverted microscope. When individual hybrid colonies are large enough to be viewed with the naked eye, transfer them with a small bore pipette to individual wells of 96 well microtiter plates and allow them to grow to confluence. At this stage, assay the supernatants with an enzyme immunoassay (EIA).

Figures 1-8 and 1-9 show the hybrid yield for fusion of G8 $\times$ SP2/0 and of B cells $\times$ SP2/0, respectively, in comparison with the yield obtained under iso-osmolar electrofusion.

METHOD III: *Iso-osmolar Electrofusion of Osmotically Prestressed Cells*

This method implies two treatment steps prior to electric field application: (1) washing and preincubating the cells in the strongly hypo-osmolar solution for about 30 min and (2) resuspension of the cells in iso-osmolar solutions and electrofusion after about 10 min. Under these conditions, hybrid yields can be obtained that are even higher than

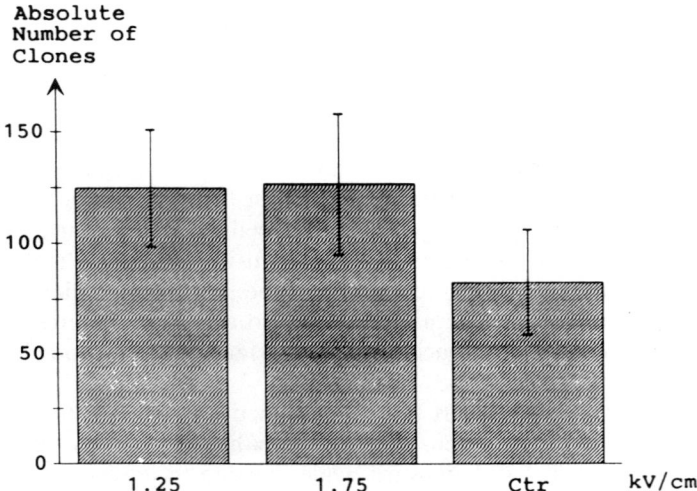

Fig. 1-8. Absolute number of hybrids obtained after electrofusion of mouse G8 hybridoma with mouse SP2/0 myeloma cells in 75 mOsmol fusion medium. The ratio of the cell densities was 1:1 and the total suspension density 2×10^5 cells per helical chamber. Field conditions: alignment field 250–300 V/cm strength and 2 MHz frequency for 30 s, followed by three fusion pulses (at intervals of 1 s) of 1.25 or 1.75 kV/cm field strength, and of 15 μs duration. The columns and error bars represent the means of six to nine experiments performed in different helical chambers and/or on different days. The control cells (ctr.) were washed and suspended in 300 mOsmol solutions and treated in the same manner except that the field strength was adjusted to 2.25 kV/cm because of the smaller cell volume in iso-osmolar solutions. (Redrawn from Schmitt and Zimmermann.[26])

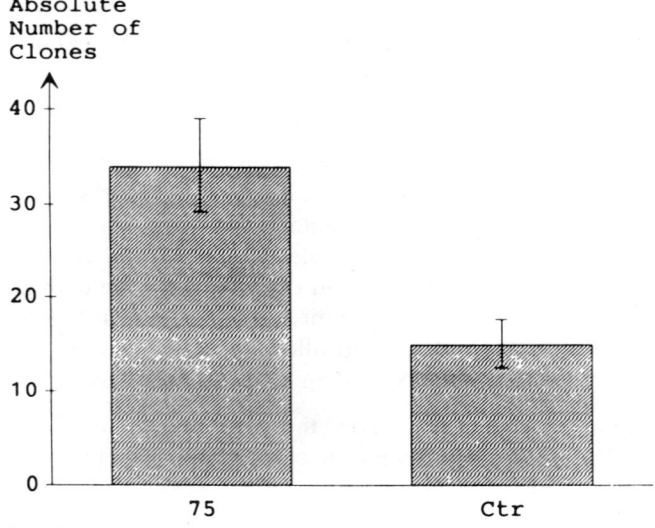

Fig. 1-9. Yield of hybridomas obtained by electrofusion of 4×10^5 mouse myeloma cells with 2×10^5 DNP-Hy-stimulated lymphocytes per helical chamber in 75 mOsmol solution (column denoted with 75). The strength of the fusion pulses was 1.5 kV/cm and the frequency of the alternating field was 1.5 MHz; otherwise the field conditions were the same as in Fig. 1-8. The control cells (ctr.) fused in iso-osmolar medium were treated in the same manner except that the field strength of the fusion pulses was adjusted to 2.00 kV/cm. The columns and error bars represent the means of four experiments performed in different helical chambers and/or on different days. (Redrawn from Schmitt and Zimmermann.[26])

those achieved with hypo-osmolar electrofusion, provided that the cells survive the two osmotic shocks in rapid succession. This is the case for the G8 × SP2/0, but not for the lymphocyte-myeloma fusion system. Application of this method requires screening experiments in which the survival of the cells is tested by counting of viable cells.

The field parameters for the fusion protocol of osmotically prestressed G8 and myeloma cells are nearly identical to those found to be optimum for iso-osmolar electrofusion of untreated cells. This is expected because volume distribution measurements of osmotically prestressed cells show that the original volume distribution of the prestressed cells is largely but not completely reestablished after transfer into iso-osmolar solution. The field conditions are: (a) alignment field of 250–300 V/cm strength, 2 MHz frequency and 30 s duration followed by three fusion pulses (1 s time interval) of 2.25 kV/cm strength and 15 μs duration, and (b) post-alignment with a constant alternating field of the same amplitude and frequency for further 30 s.

Procedure

1. Harvest and mix the cells as described under Method II
2. Wash the cell mixture twice in 75 mOsmol fusion solution by resuspending the cell pellets in 10 ml of the washing medium and subsequent spinning down for 10 min at 150 g.
3. Resuspend the pellet in 340 μl of 300 mOsmol fusion solution.
4. Fill the 340 μl into the helical chamber.
5. Apply the preset program for electrofusion.
6. Handle and treat the field-exposed cells as described under Method II.

Figure 1-10 shows the yield of hybrids obtained under these conditions in comparison to the yield of iso-osmolar electrofusion of osmotically untreated cells.

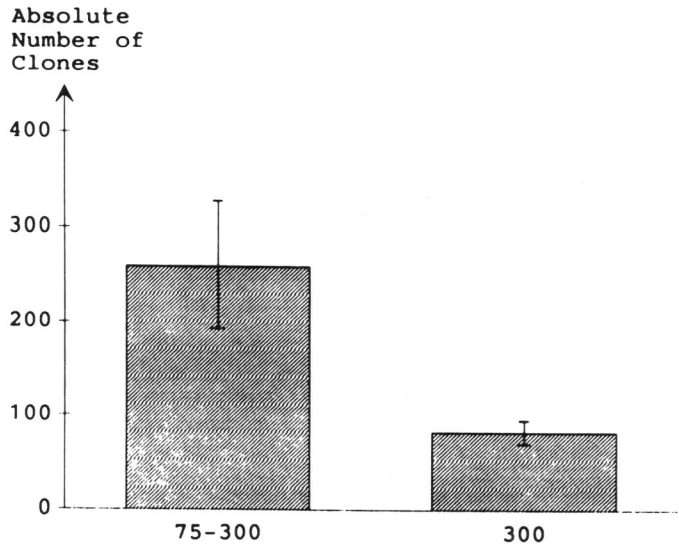

Fig. 1-10. Absolute hybrid yield obtained by electrofusion of G8 hybridoma with SP2/0 myeloma cells in iso-osmolar medium after osmotic prestress in 75 mOsmol solution (column denoted with 75-300). The column denoted with 300 represents the control fusion experiments in iso-osmolar medium without hypo-osmolar pretreatment of the cells. (Field conditions as in Fig. 1-8.) Note that the field strength was adjusted to 2.25 kV/cm. The columns and error bars represent the means of 4 to 6 experiments performed in different helical chambers and/or on different days. (Redrawn from Schmitt et al.[25])

V. POINTS TO CONSIDER

The key step in electroinjection and electrofusion is the reversible electrical breakdown of the cell membrane. Electrical breakdown is achieved when freely suspended cells, cell pairs or higher aggregates, or cell layers are exposed to a field pulse of high intensity (kV/cm) and short duration (microseconds). The optimum field conditions have to be carefully selected for each cell fusion or injection system. In most cases, failure in electroinjection and -fusion arises from the choice of the wrong field conditions. The following considerations are meant as a guide to find the optimum parameters.

A. Calculation of the Field Strength of the Breakdown Pulse for Spherical Cells

The generated membrane potential across planar membranes can be calculated by dividing the external field strength by the membrane thickness, provided that the external solutions facing the membrane are highly conductive compared to the membrane. However, physical theory shows that this is not the case if we are dealing with cells, that is, particles containing a very conductive cell interior and surrounded by a nearly insulating shell (membrane). In the case of spherical cells the generated membrane potential, which is superimposed on the intrinsic membrane potential (see below), must be calculated using the following equation (Fig. 1-1[33]):

$$V_c = f\, a\, E_c \cos \alpha \qquad (1\text{-}1)$$

where V_c = breakdown voltage, a = cell radius, E_c = critical field strength, α = angle between a given membrane site and field direction, and f is the shape factor that assumes a value of 1.5 for spherical cells.

Equation 1-1 shows that for a spherical cell, the generated membrane potential is radius- and angle-dependent. This has important consequences for electroinjection (and also for electrofusion). First, the external field strength, E_c, which is required to reach the breakdown voltage of the cell membrane, V_c, depends on the radius of the cell. Second, for a given cell size and field strength, the distribution of the generated membrane potentials is not equal over the entire membrane surface. For membrane sites oriented in field direction (i.e., "the poles" of freely suspended or aligned cells), the angle α is zero and therefore cosine $\alpha = 1$ (Eq. 1-1). The generated membrane potential is always highest at these sites. At membrane sites oriented perpendicular to the field lines ("equator," $\alpha = 90°$), cosine 90° is equal to zero. Therefore, the generated membrane potential is zero, independent of the field intensity applied across the cell suspension. Breakdown will never occur at these sites.

Between the "poles" and the "equator" the generated membrane potential decreases progressively and is determined by the magnitude of cosine α. The consequence of this is that the breakdown voltage is first reached in field direction and then—at higher field strengths—at membrane sites oriented to the field lines at a certain angle α where $0 < \alpha < 90°$.

The increase in permeability induced by breakdown is therefore also a function of the field strengths. At the critical field strength, permeabilization occurs only at the "poles" (or in the contact zone of aligned cells) and at critical field strengths over larger membrane areas determined by the cosine term in Eq. 1-1. Thus, we can match the field-induced permeability change to the size (and charge) of the molecule that is injected. For aligned cells, the breakdown area determines the rate of fusion.

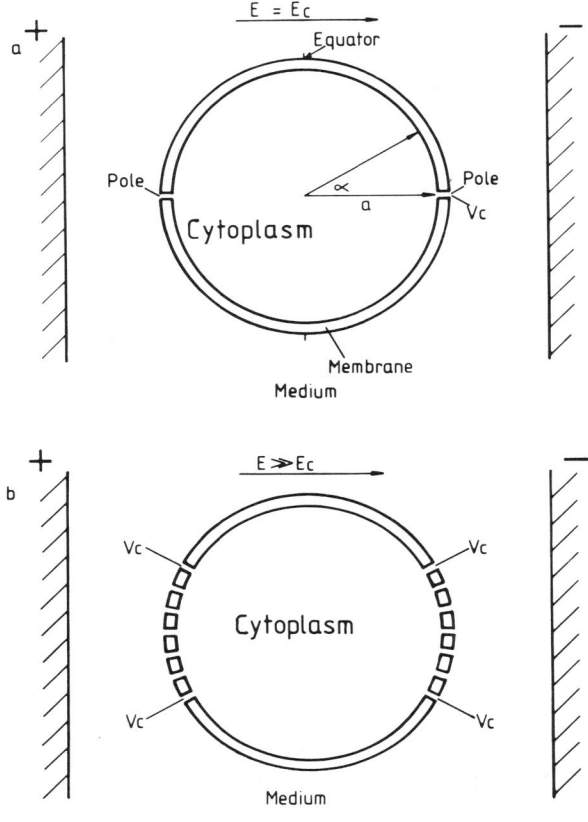

$$V_c = 1{,}5 \cdot a \cdot E \cdot \cos \alpha$$

Fig. 1-11. Schematic diagram of a spherical cell exposed to a field pulse of high intensity. As indicated by the equation the generated potential is radius- and angle-dependent. The breakdown voltage, V_c, is therefore first reached in field direction for a critical field strength of $E = E_c$. The breakdown voltage of membrane sites oriented at a certain angle α to the field direction is reached only if supracritical field strengths ($E > E_c$) are applied. Breakdown of local areas of the cell membrane is indicated by the formation of transmembrane pores.

In order to establish the optimum field strengths for electroinjection and -fusion, we have to measure the volume distribution of the cells (or at least the modal volume of the cell population) by viewing under the microscope or—more accurately—by a particle volume analyzer (Coulter counter[34]). This is very important if the cells are fused in hypo-osmolar solutions. In these solutions, cells swell considerably, and therefore the field strength of the breakdown pulse for injection or fusion has to be reduced correspondingly.

Figure 1-11 represents only a simplified view of the events that occur in the membrane at critical and supracritical field strengths. In reality, the pole areas of the cells will be subjected to larger perturbations than other areas if supracritical field strengths are applied, as is the case in protein and DNA injection. This may lead to irreversible changes in the membrane structure of these areas, resulting in cell death. However, this problem can be overcome by correct interfacing of the electronics and the pulse medium (see "Methods").

Equation 1-1 can be used for calculation of the field strength required for electroinjection (and also for electrofusion). Figure 1-2 represents the critical field strengths (which are required for breakdown at the cell poles, cos 0° = 1) versus the radius of typical cells. The curves were calculated for two breakdown voltages of 1 V (room temperature) and 2 V (4°C).

It is obvious that the required field strength decreases in the following order: bacteria (1 μm) > yeast cells (3 μm) > mammalian cells (10 μm) > plant protoplasts (20 μm) > eggs (200 μm). If macromolecules are injected, it is not sufficient to induce only breakdown restricted to the pole areas of the cells. Larger areas have to be subjected to breakdown. In this case the critical field strength calculated according to Eq. 1-1 has to be multiplied by a factor of 5 to 7 in order to take into account the angular dependence of permeabilization.

In the case of cells exhibiting an elongated (ellipsoid) shape, Eq. 1-1 is no longer valid. The mathematics for the calculation of the generated membrane potential do not lead to a simple expression. Only for the "pole areas" of the cells can a simple equation be deduced (by analogy to Eq. 1-1):

$$V_c = f\, a\, E_c \qquad\qquad (1\text{-}2)$$

In this equation, the shape factor f assumes a value of 1. In practice, however, we can use the rule mentioned earlier, that multiplication of the critical field strength calculated according to Eq. 1-2 by a factor of about 5 will lead to a permeabilization of the membrane sufficient for the uptake of macromolecules. For cell shapes between spheres and ellipsoids, factors have to be used that are larger than 1 and smaller than 1.5. For spherical cells that do not have a smooth surface, the situation is more difficult from a theoretical standpoint. In this case (e.g., when using macrophages), Eq. 1-1 should be used as a guideline for a rough estimation of the generated membrane potential. Then, the optimum field strength must be found by trial and error using, for example, dye uptake as a criterion of optimum permeabilization. However, it is extremely important to note that Eqs. 1-1 and 1-2 or the curves in Fig. 1-12 can only be used if the stationary state is reached. This means that it is assumed in the calculations that the duration of the pulse is long enough so that the final voltage given by the equations is reached. This is not

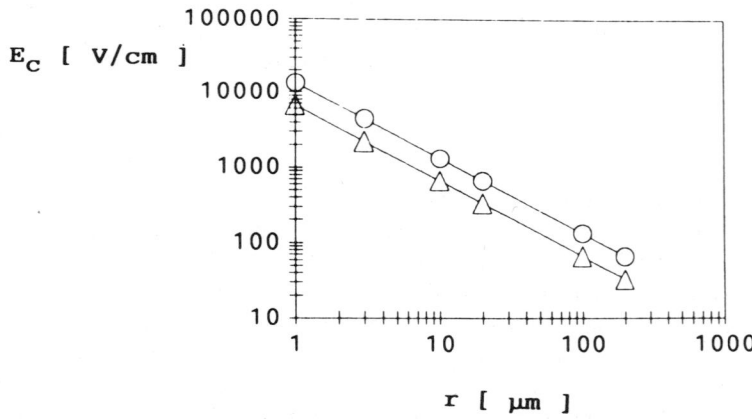

Fig. 1-12. The critical field strength (required for breakdown of the cell membrane in field direction, see Fig. 1-11) calculated as a function of the cell radius according to Eq. 1-1 (see "Points to Consider"). Calculations were made by assuming a membrane breakdown voltage of 1 V (room temperature, triangles) and of 2 V (4°C, circles).

always the case, and is one of the main reasons why electroinjection (and electrofusion) may not lead to high yields of transformants (or hybrids).

B. Membrane Charging and the Role of Medium Conductivity

If a field is applied, the potential will not be instantaneously built up across the membrane. Ion movement and therefore charging of the membrane takes time. In a circuit consisting of a capacitor and a resistance arranged in parallel to each other, the voltage across the capacitor increases in an exponential manner in response to the external field. An exponential process is characterized by a constant, the so-called relaxation time, τ. The relaxation time is the time that is needed to reach 63% of the final value (given by Eqs. 1-1 or 1-2).

The final equilibrium potential is reached after about 5 times the relaxation time. If this period is shorter than the duration of the applied field pulse, we can use Eq. 1-1 and Fig. 1-12 for calculation of the critical field strengths without any further corrections.

The relaxation time, τ, of a spherical cell can be calculated from the following equation:

$$1/\tau = 1/R_m C_m + 1/a C_m \, (\rho_i + 0.5\rho_e) \tag{1-3}$$

where R_m = specific membrane resistance (Ωcm^2); C_m = specific membrane capacitance ($\mu F/\text{cm}^2$); ρ_i = internal specific resistance (Ωcm) and ρ_e = external specific resistance (Ωcm). In electroinjection (but also in electrofusion for a first approximation), the resistance of the membrane of untreated cells, R_m, is (with very few exceptions) always higher than that of the external solution. Thus, the first term in Eq. 1-3 is negligible compared with the second term. Therefore, in most cases it is adequate to use the following approximation:

$$\tau = a \, C_m \, (\rho_i + 0.5 \, \rho_e) \tag{1-4}$$

Equation 1-4 shows that the relaxation time is a function of both the radius of the cell and of the external conductivity. The membrane capacitance and the internal conductivity can be assumed to be constant for most living cells. Their values are $C_m = 1 \, \mu F/\text{cm}^2$ and $\rho_i = 100 \, \Omega\text{cm}$.

Figure 1-13 represents the relaxation time as a function of the cell radius for two external conductivities of the hypo-osmolar media (pulse and fusion medium), and Fig. 1-14 shows the relaxation time as a function of the external conductivity, ρ_e, for different cell radii. The range of external conductivity and cell radii covers the range that is normally used in electroinjection (and in electrofusion).

It is obvious that changes in the electrolyte composition of the pulse or fusion medium may considerably alter the field conditions. In each case, it is necessary to adapt the field parameters to the new environment according to Eq. 1-1 (Fig. 1-12) and to Eq. 1-4 (Figs. 1-13 and 1-14) This can be performed either by changes of the pulse duration or of the field strength.

C. Temperature

The temperature is a crucial parameter that must be controlled carefully during the entire process of electroinjection. The breakdown voltage is a strong function of temperature, as has been shown by direct measurements on giant algal cells and planar lipid bilayer membranes.[35,36] In these systems, the breakdown voltage can be measured directly with

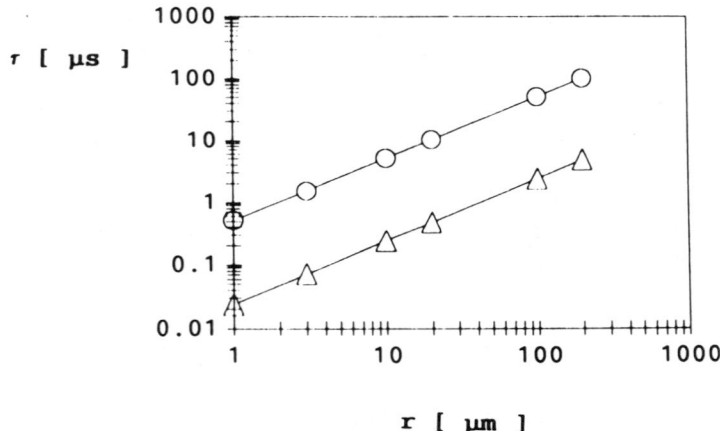

Fig. 1-13. The relaxation time of the charging process of the membrane in response to a field pulse as a function of the cell radius. The curves are calculated from Eq. 1-4 for two different external resistivities, ρ_e (triangles, 286 Ωcm, and circles, 10^4 Ωcm) corresponding to the conductivities of the hypo-osmolar pulse and fusion media, respectively. It was assumed that the specific membrane capacitance, C_m, is 1 $\mu F/cm^2$ and the internal resistivity, ρ_i, is 100 Ωcm.

an internal microelectrode. These investigations, as well as indirect studies on cells (by using a special particle analyzer), have shown that the breakdown voltage of living cells is about 2 V at 4°C, 1 V at room temperature (22°C), and 0.5 V at 37°C.

Electroinjection work performed in various laboratories for more than 15 years suggests that the highest incorporation of foreign molecules (particularly of macromolecules) can be achieved if pulsing occurs at about 4°C. (It should be noted, however, that yeast cells may represent an exception.) The reason for this is obvious. The resealing and therefore the lifetimes of the field-induced perturbations in the membrane are dependent

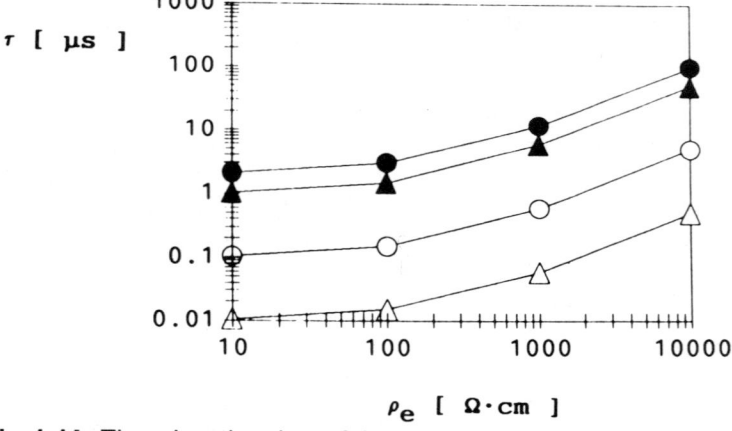

Fig. 1-14. The relaxation time of the charging process of the membrane in response to a field pulse as a function of the external resistivity, ρ_e, for different cell radii (open triangles, 1 μm; open circles, 10 μm; closed triangles, 100 μm; and closed circles, 200 μm). For calculations and assumptions, see Fig. 1-13.

on temperature. Between 25° and 37°C, resealing may be complete within a couple of seconds to 1 min (depending on the species). In contrast, at low temperatures the high permeability state of the membrane may be maintained for 10 min or more because of the slow reversal of field-induced conformational changes in membrane proteins. However, uptake of foreign molecules, particles, organelles, and smaller cells into host cells occurs only by diffusion, according to the concentration gradients existing between the cell and the environment. If the membrane reseals too fast (due to use of higher temperatures), either small amounts of the foreign molecules are taken up or large concentrations of the compound to be incorporated have to be added to the external solution in order to provide sufficient uptake. This is unfavorable, particularly if the entrapment of DNA is considered. In most cases only a small amount of plasmid is available. Performance of electroinjection at 4° to 6°C is therefore strongly recommended. However, several cell system parameters have to be considered for successful electroinjection.

First, many cells (particularly mammalian) can only be kept in iso-osmolar and hypo-osmolar solutions for a certain time at these low temperatures without irreversible changes to the cellular and membrane functions. The best compromise between uptake and viability is normally achieved by precooling of the chamber and by pulse application immediately after introduction of the cell suspension and connection of the chamber to the apparatus. Once the pulse or pulses have been applied, it is very important that the pulsed suspension is kept for an additional 2 to 3 min at low temperature, followed by an increase of the temperature to 25°C (plant protoplasts, eggs), 30°C (yeast cells), or to 37°C (mammalian and bacterial cells) in order to entrap the incorporated substances within the cells. The cells must be kept for an additional 30- to 60-min period at these elevated temperatures. Otherwise, the restoration process is not fully completed.

During the period of temperature increase, it is very important that any vigorous shaking or suction is avoided, particularly if mammalian and plant cells or enzyme-treated eggs are used. The permeabilized cells are very sensitive to shear forces arising from such operations.

Experience has shown that gentle withdrawal of the pulsed cell suspension with a micropipette (having a broad tip) and immediate reinjection in warmed post-pulse medium is the best way to achieve a high yield of intact, manipulated cells. For electrofusion, the temperature should be adjusted to about 25°C. This temperature allows rapid resealing of the breakdown area at the poles of terminal cells in a chain and, therefore, minimizes release of intracellular substances. Conversely, this temperature is not too high to prevent the intermingling process of the membranes of attached cells. This can happen at still higher temperatures, however, resulting in a reduction of hybrid yield.

D. Pulse Duration

The duration of the applied field pulse(s) is most critical and ignorance of the physics behind this often leads to failure in electroinjection and -fusion. Before breakdown, the cell interior is shielded against the external field by the membrane. After localized breakdown, field lines will pass partly through the cytoplasm and high current densities will occur locally within the membrane and the cell. Consequently, secondary processes such as local heating, electrophoresis of membrane and cytoplasm, and molecule-pronounced osmotic processes come into play. These factors are ultimately lethal for the cells. Breakdown is a rapid event; it occurs in less than 50 ns once the critical voltage across the membrane is reached. Thus, the field exposure time after breakdown must be kept to a minimum to avoid current flow through the conductive cell interior. The guidelines for adjustment of the pulse length to an optimum value are given by Figs. 1-12 through 1-14. First, one must calculate the time that is needed to reach the breakdown voltage at a given medium conductivity and cell radius. This means that the relaxation time must be deduced from Figs. 1-13 and 1-14, respectively, and multiplied by a factor of 5 as mentioned earlier. Then, the field strength must be calculated from Eqs. 1-1 or 1-2 (or from Fig. 1-12) and possibly corrected by taking the "breakdown time" into account.

Generally, the pulse length should not be larger by a factor of 2 to 3 than this calculated "breakdown time," if reversible breakdown and high efficiency of electroinjection and -fusion are desired. It is obvious from the calculations in Figs. 1-13 and 1-14 that the pulse duration should be between 5 and 15 μs when mammalian cells suspended in conductive (i.e., electrolyte-containing) media are permeabilized. For plant protoplasts, somewhat larger pulse durations can be applied without detrimental effects on the majority of the cells. In the extreme case of eggs that may have a radius of 200 μm to 1 mm, the pulse duration can be extended to some ms because of the very long relaxation time. In the opposite extreme of bacteria, the pulse duration theoretically should be less than 1 μs in the light of the discussion here. Because of the absence of compartmentalization, however, bacteria survive longer pulse durations as is the case with human or bovine erythrocytes.

E. Pulse Trains

Electroinjection of macromolecules in iso-osmolar (but not in hypo-osmolar) media often requires the injection of several pulses of moderate, supracritical amplitude (critical field strength multiplied by a factor of about 5, see above) in order to achieve high yields of incorporation.[31] It is very important, however, that the time interval between two consecutive pulses is of the right order, because otherwise such a pulse sequence would lead to irreversible side effects. This important point is very often overlooked. Many researchers inject pulse trains with a time interval of 1 s between consecutive pulses. Even though this procedure is favorable in electrofusion of aligned cells, it is lethal to freely suspended cells. As mentioned earlier, current will flow through the permeabilized membrane area and through the conductive cell interior. Conversely, resealing—particularly at low temperatures—requires more time. Without resealing of the bilayer, a membrane potential could not be built up across the membrane during the consecutive pulses. The only event that occurs in this procedure is that current flows through the cell, leading to all of the detrimental side effects discussed previously.

Therefore, it is an absolute necessity to increase the time interval between consecutive pulses to at least 30 to 60 s in order to allow sufficient resealing of the lipid bilayer structure at low temperature. Resealing of the bilayer still proceeds at low temperatures, whereas protein restoration requires many minutes. If the time interval between consecutive pulses is adjusted to about 1 min, the application of a pulse train has the advantage that the uptake of macromolecules can be greatly increased. This is because of slight rotation and movement of the cells induced by Brownian motion, sedimentation, and convection. Membrane sites, which were not (or only slightly) exposed to the preceding pulse (see Fig. 1-11), will be oriented in the direction of the field. In this way, the problem of a restricted membrane area receiving the maximum potential is circumvented.

Simultaneously there is an increased likelihood that macromolecules such as DNA, which are absorbed to the external membrane surface, are taken up in large amounts. If these boundary conditions are taken into account, electroinjection will lead to reasonably high injection yields in iso-osmolar solutions.

In electrofusion, several pulses must be applied at intervals of 1 s. In this method, the secondary effects that follow the first breakdown of the membrane are used to enlarge the breakdown area in the contact zone and to facilitate the formation of a cytoplasmic continuum between the two attached cells (see also Section F, "Asymmetric Breakdown and Media Composition"). A cytoplasmic continuum prevents the resealing process of the individual membranes in the contact zone and thus stimulates the intermingling process of the membranes of the two attached cells.

F. Asymmetric Breakdown and Media Composition

As mentioned earlier, the electrical breakdown of the cell membrane is governed by the total potential across the membrane, which is made up of the intrinsic membrane potential and that generated by the field. Figure 1-11 addresses only the distribution of the generated potential and not the intrinsic membrane field. As discussed earlier, however, if the intrinsic membrane potential (or field) is relatively high, asymmetric breakdown (restricted to only one hemisphere of the cell) may occur. This is because the two potential differences are superimposed in parallel in one hemisphere, but in antiparallel in the other (Fig. 1-15). Consequently, if critical field strengths are applied, breakdown will occur only in the hemisphere of the membrane oriented to the anode. It can be assumed that the cell interior is negatively charged with respect to the external solution. Asymmetric breakdown (restricted to one hemisphere) is advantageous in electroinjection. The current that will enter the cell is limited by the intact (high) resistance of the opposite, nonpermeabilized hemisphere. Thus, the cell is shielded against

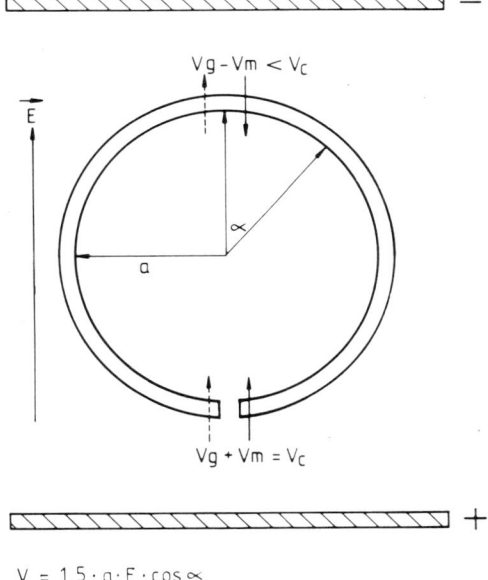

$$V = 1.5 \cdot a \cdot E \cdot \cos \alpha$$

Fig. 1-15. Schematic diagram of a cell exposed to a uniform field pulse, E, as in Fig. 1-11, showing how the membrane potential, V_g, generated across the membrane, is superimposed on the intrinsic membrane potential, V_m. It is assumed that the cell interior is negatively charged with respect to the external medium; therefore the vector of the intrinsic field is directed from the outside to the inside of the cell. Note that in the case of a field pulse of critical field strength, the breakdown voltage (which is equal to $V_m + V_g$) is reached only in the membrane of the hemisphere that faces the anode (positive electrode).

adverse side effects arising from the field, even though the membrane is permeabilized in one hemisphere.

By using fluorescent dyes as probe molecules, it has been demonstrated that only the hemisphere oriented to the anode is permeabilized. This was demonstrated both for plant protoplasts and mammalian cells.[38,39] The production of stable transformants in DNA-electroinjection shown in Figs. 1-5 through 1-7 was performed under these conditions. The occurrence of asymmetric breakdown depends on the magnitude of the intrinsic field, which itself is controlled—among other things—by the surface charges and the corresponding surface potentials. Changes in the surface potentials can be induced by changes of the ionic strength of the external medium, for example, by addition of multivalent ions (e.g., calcium) or by pretreatment of the cells with digestive enzymes. Due to these procedures, it is possible that asymmetric breakdown becomes symmetrical or even reverses to give an asymmetric breakdown of the opposite hemisphere (oriented to the cathode). The consequences of this are that if the ionic composition of the medium is changed because of unfavorable values of the charging time of the membrane (see above), the intrinsic field can change in a way that may prevent asymmetric breakdown. In addition, the calculations of the breakdown voltage according to Eqs. 1-1 or 1-2 will be erroneous in the presence of high intrinsic fields, because these equations assume that the intrinsic membrane potential is zero.

Asymmetric breakdown also occurs in electrofusion of cells because of the low ionic strength of the fusion medium. Therefore, supracritical field strengths and several pulses within a short interval must be applied to permeabilize the membranes of both cells at the same time. Preswelling of the cells in hypo-osmolar solutions apparently facilitates permeabilization of both membranes in the contact zone because of the dissolution of the cytoskeleton.

G. Alignment

In the presence of an alternating field, a membrane potential is generated that is given by Eq. 1-5[40]:

$$V = \frac{1.5\, a\, E\, \cos\alpha}{\sqrt{1 + (\omega\tau)^2}} \tag{1-5}$$

where ω is the angular frequency of the alternating field. Equation 1-5 takes into account the frequency dependence of the generated membrane potential (see Eq. 1-1). In practice, the generated membrane potential is frequency-independent up to 1 MHz in order to decrease linearly above this threshold value. At these frequencies, the membrane capacitor becomes short-circuited and current will flow partly through the conductive cell interior. This has two consequences. Higher field strengths must be applied to the cells in order to reach the breakdown voltage of the plasmalemma membrane, and an inhomogeneous electric field is generated within the cells, which leads to internal dielectrophoresis of organelles. In certain very narrow frequency ranges above 1 MHz, it is therefore possible to induce migration of nuclei into the contact zone of adhered cells.[41] This results in fusion of the plasmalemma and in fusion of the nuclear membranes of apposed nuclei, provided that sufficient field strengths are applied to break down the membranes. The dielectrophoretic force, F_D, which acts on the cell in an inhomogeneous alternating electric field, is given by Eq. 1-6[42]:

$$F_D = 2\,\pi\, a^3\, \epsilon_0\, \epsilon_1\, \frac{\epsilon_2 - \epsilon_1}{\epsilon_2 + 2\,\epsilon_1}\, \nabla E^2 \tag{1-6}$$

where ϵ_1 is the relative dielectric constant of the medium and ϵ_2 that of the cell; ϵ_O is the dielectric constant of vacuum; ∇ is the Nabla operator and is an indicator for the inhomogeneity of the field.

Equation 1-6 indicates that the dielectrophoretic force is proportional to the volume of the cells. This results in cell separation if the two fusion partners have different volumes, as is the case for lymphocytes and myeloma cells both in iso- and hypo-osmolar solutions. The larger (myeloma) cells move faster toward the electrodes than the smaller (lymphocyte) cells. Thus, alignment of myeloma cells will occur first, which is not desirable because of the reduction of hybrid yield. The effect of cell separation due to the volume dependence of the dielectrophoretic force can be avoided if the amplitude of the alternating field is first increased to very high values and then reduced by a ramp to smaller values, which are just sufficient to keep the aligned cells in position. At very high field strengths of the alternating field, the dielectrophoretic force is so strong that cell chain formation will occur in a few seconds without separation of the two differently sized cell types. Subsequent reduction of the field strength to low values then avoids irreversible changes in the membrane.[18]

Inspection of Eq. 1-6 also shows that the cells will only move to the electrodes (region of highest field intensities) when the dielectric constant of the cell is higher than that of the medium (positive dielectrophoresis). In the opposite case, cell alignment will occur in the middle of the electrode gap or the cells will move out of the electrode space (region of weakest field intensity, so-called negative dielectrophoresis). Since the dielectric constants are functions of the frequency of the alternating field, frequency regions are observed in which positive or negative dielectrophoresis occur.

Equation 1-6 holds only for a dielectric body, not for cells. Therefore, the equation is only a tool for the qualitative estimation of the frequency dependence of the cell behavior during dielectrophoresis. However, it is sometimes very difficult to predict the region of positive dielectrophoresis, which is normally used in electrofusion. The best way to find the region of optimum positive dielectrophoresis is the use of field probe particles (small bacteria or polystyrene particles). These particles align along the field lines and are efficient indicators for the field distribution around aligned (larger) cells.[43]

VI. CONCLUSIONS

The foregoing considerations make it apparent that a combination of interdependent cell and system parameters determines the success of electroinjection and electrofusion. Physical theory predicts that any compound can be injected into any given cell type in high yield and that any cell system can be fused provided that we keep all these boundary conditions in mind. In practice, however, the situation is more simple, as is clear from the previous detailed discussion. This is particularly true when electroinjection and -fusion are performed in strongly hypo-osmolar solutions with electronic equipment and media, which are adapted to the cell parameters. Obviously, the osmotic stress and the associated tensions created in the membrane plane of the swollen cells change the membrane structure (and permeability), which apparently facilitates electroinjection and -fusion. However, because no significant effects on membrane transport properties and cell viability are observed, it must be assumed that these changes in membrane structure are minor. Such small changes in membrane structure can be envisaged if the membrane and/or cytoskeletal proteins are (partly) dissolved by the osmotic stress, resulting in a higher mobility of membrane components and probably in the emergence of protein-free lipid domains. The experiments reported here force the conclusion that disruption of restraining proteins by osmotic means is more advantageous than by field pulses alone.

Fusion yields are very high and reproducible and more independent of culture conditions and of the number of cell passages. This is in contrast to the experience gathered using electroinjection and -fusion in iso-osmolar solutions. Under the latter conditions, optimum field conditions often change once the cells have been passaged several times or when the ingredients of the culture medium are changed. The reason for this variability in fusion yield is evident from the explanation given earlier. Slight changes in the cell and membrane skeleton proteins during growth and passage of cells (e.g., induced by overgrowth of a subpopulation) can alter the interaction of the field vector with the membrane components of the freely suspended or dielectrophoretically aligned cells.

The present protocols for formation of transformants and hybrids also demonstrate that osmotic pressure gradients and associated tensions in the membrane do not play a primary role in the initiation of the electrofusion process. This contradicts the suggestion of Ahkong and Lucy.[44,45] Otherwise, it cannot be explained why fusion of osmotically prestressed cells in iso-osmolar solutions (see Method III) yields even higher numbers of hybrids than in iso- or hypo-osmolar solutions without osmotic pretreatment (although the volume distribution of the cells is nearly reestablished).

Further elucidation of the mechanism that facilitates plasmid uptake and fusion of osmotically stressed cells can be envisaged through the use of plant cells. These cells are large enough to allow the study of the spatial permeabilization induced by the combined effects of osmotic pressure and electric field pulses with fluorescently labeled compounds and chemotactic bacteria.[39]

ACKNOWLEDGMENTS

The authors are very grateful to J. J. Schmitt and R. Däumler for preparing the illustrations. This work was supported by grants for the Federal Ministry of Research and Technology (BMFT) of the Fondorf Chemical Industry and the Deutsche Forschungsgemeinschaft (SFB 176) and by grants HL33811, AI22557 and AI26031 from the National Institutes of Health (U.S.A.).

REFERENCES

1. Zimmermann, U., Schulz, J., and Pilwat, G. 1973. Transcellular ion flow in *E. coli* B and electrical sizing of bacterias, Biophys. J. 13:1005–1013.
2. Zimmermann U., Pilwat, G., and Riemann, F. 1974. Patent for electroinjection of macromolecules into living cells, filed in Feb. 1974, no. 2405119 (F.R.G.), no. 1481480 (U.K.), no. 4081340 (USA) and other countries.
3. Zimmermann, U., Pilwat, G., and Riemann, F. 1974. Reversible dielectric breakdown of cell membranes by electrostatic fields. Z. Naturforsch. 29c:304–310.
4. Zimmermann, U., Riemann, F., and Pilwat, G. 1976. Enzyme loading of electrically homogeneous human red blood cell ghosts prepared by dielectric breakdown. Biochim. Biophys. Acta 436:460–474.
5. Zimmerman, U., Pilwat, G. and Esser, B. 1978. The effect on encapsulation in red blood cells on the distribution of methotrexate in mice. J. Clin. Chem. Clin. Biochem. 16:135–141.
6. Vienken, J., Jeltsch, E., and Zimmermann, U. 1978. Penetration and entrapment of large particles in erythrocytes by electrical breakdown techniques. Cytobiology 17:182–198.
7. Zimmermann, U. 1983. Cellular drug carrier systems and their possible targeting. In Targeted drugs. E. P. Goldberg, ed., John Wiley-Verlag, New York. pp. 153–200.
8. Auer, D., Brandner, G., and Bodemer, W. 1976. Dielectric breakdown of the red blood cell membrane and uptake of SV 40 DNA and mammalian cell RNA. Naturwissenschaften 63 (8):391.

9. Schüssler, W., and Ruhenstroth-Bauer, G. 1984. Stomatocytosis of Latex particles (0.26 μm) by rat erythocytes by the electrical breakdown technique. Blut 49:213–217.
10. Neumann, E., Schaeffer-Ridder, M., Wang, Y., and Hofschneider, P. H. 1982. Gene transfer into mouse lyoma cells by electroporation in high electric field. EMBO J. 1:841–845.
11. Calvin, N. M., and Hanawalt, P. C. 1988. High-efficiency transformation of bacterial cells by electroporation. J. Bacteriol. 170 (6):2796–2801.
12. Hibi, T., Kano, H., Sugiura, M., Katamu, T., and Kimura, S. 1986. High efficiency electrotransfection of tobacco mesophyll protoplasts with TMV RNA. J. Gen. Virol. 67:2037–2042.
13. Hama-Inaba, H., Shiomi, T., Sato, K., Ito, A., and Kasai, M. 1986. Electric pulse-mediated gene transfer in mammalian cells grown in suspension culture. Cell Struct. Funct. 11:191–197.
14. Stopper, H., Zimmermann, U., and Neil, G. A. 1988. Increased efficiency of transfection of murine hybridoma cells with DNA by electropermeabilization. J. Immun. Meth. 109:145–151.
15. Zimmermann, U., and Pilwat, G. 1978. The relevance of electric field induced changes in the membrane structure to basic membrane research and clinical therapeutics and diagnosis. In: Abstracts of the sixth International Biophysics Congress, Kyoto, Japan, p. 140.
16. Zimmermann, U., and Scheurich, P. 1981. Fusion of Avena sativa mesophyll cell protoplasts by electrical breakdown. Biochim. Biophys. Acta 641:160–165.
17. Zimmermann, U., Scheurich, P., Pilwat, G., and Benz, R. 1981. Cells with manipulated functions: New perspectives for cell-biology, medicine and technology. Angew. Chemie, Int. Ed. Engl. 20:325–344.
18. Zimmermann, U. 1982. Electric field-mediated fusion and related electrical phenomena. Biochim. Biophys. Acta 694:227–277.
19. Zimmermann, U. 1983. Electrofusion of cells: Principles and industrial potential. Trends Biotechnol. 1:149–155.
20. Zimmermann, U. 1986. Electrical breakdown, electropermeabilization and electrofusion. Rev. Physiol. Biochem. Pharmacol. 105:175–256.
21. Zimmermann, U. Electrofusion of cells. In Methods of Hybridoma Formation. A. H. Bartal, Y. Hirshaut, eds., Humana Press, New York, 1987.
22. Zimmermann, U., Urnovitz, H. B. 1987. Principles of electrofusion and electropermeabilization. Meth. Enzym. 151:194–221.
23. Zimmermann, U. Electrofusion and electrotransfection of cells. In Molecular Mechanisms of Membrane Fusion, S. Ohki, D. Doyle, T. D. Flanagan, S. W. Hui, E. Mayhew, eds., Plenum Press, New York and London, 1988.
24. Zimmermann, U., Schmitt, J. J., and Kleinhans, P. The potential of electrofusion for hybridoma production. In Clinical Applications of Monoclonal Antibodies, R. Hubbard and V. Marks, eds., Plenum Press, New York and London, 1988.
25. Schmitt, J. J., Zimmermann, U., and Gessner, P. 1989. Electrofusion of osmotically treated cells: high and reproducible yields of hybridoma cells. Naturwissenschaften 76:122–123.
26. Schmitt, J. J., and Zimmermann, U. 1989. Enhanced hybridoma production by electrofusion in strongly hypo-osmolar solutions. Biochim. Biophys. Acta 938:47–50.
27. Däumler, R., and Zimmermann, U. 1989. High yields of stable transformants by hypo-osmolar plasmid electroinjection. J. Immun. Meth. 122:203–210.
28. Chapter 3.
29. Southern, P. J., and Berg, P. 1982. Transformation of mammalian cells to antibiotic resistance with a bacterial gene under control of the SV 40 early region promoter. J. Mol. Appl. Gen. 1:327–341.
30. Stopper, H., Zimmermann, U., and Wecker, E. 1985. High yields of DNA-transfer into mouse L-cells by electropermeabilisation. Z. Naturforsch. 40c:929–932
31. Stopper, H., Jones, H., and Zimmermann, U. 1987. Large scale transfection of mouse L-cells by electropermeabilisation. Biochim. Biophys. Acta 900:38–44.
32. Schmitt, J. J., Zimmermann, U., and Neil, G. 1989. Efficient generation of stable antibody forming hybridoma cells by electrofusion. Hybridoma 8:107–115.
33. Jeltsch, E., and Zimmermann, U. 1979. Particles in a homogeneous electrical field: A model for the electrical breakdown of living cells in a Coulter Counter. Bioelectrochem. Bioenerg. 6:349–384.
34. Broda, H.-G., Schnettler, R., and Zimmermann, U. 1987. Parameters controlling yeast hybrid yield in electrofusion: The relevance of pre-incubation and skewness of the size distributions of both fusion partners. Biochim. Biophys. Acta 899:25–34.

35. Coster, H. G. L., and Zimmermann, U. 1975. The mechanism of electrical breakdown in the membranes of *Valonia utricularis*. J. Membrane Biol. 22:73–90.
36. Benz, R., and Zimmermann, U. 1981. The resealing process of lipid bilayers after reversible electrical breakdown. Biochim. Biophys. Acta 640:169–178.
37. Benz, R., and Zimmermann, U. 1980. Pulse-length dependence of the electrical breakdown in lipid bilayer membranes. Biochim. Biophys. Acta 597:637–642.
38. Mehrle, W., Zimmermann, U., and Hampp, R. 1985. Evidence for asymmetrical uptake of fluorescent dyes through electro-permeabilised membranes of *Avena* mesophyll protoplasts. FEBS Lett. 185:89–94.
39. Mehrle, W., Hampp, R., and Zimmermann, U. 1989. Electric pulse induced membrane permeabilisation. Spatial orientation and kinetics of solute efflux in freely suspended and dielectrophoretically aligned plant mesophyll protoplasts. Biochim. Biophys. Acta 978:267–275.
40. Holzapfel, C., Vienken, J., and Zimmermann, U. 1982. Rotation of cells in an alternating field: Theory and experimental proof. J. Membr. Biol. 67:13–26.
41. Bertsche, U., Mader, A., and Zimmermann, U. 1988. Nuclear membrane fusion in electrofused mammalian cells. Biochim. Biophys. Acta 939:509–522.
42. Pohl, H. A. Dielectrophoresis. Cambridge University Press, Cambridge, 1978.
43. Mehrle, W., Hampp, R., Zimmermann, U., and Schwan, H. P. 1988. Mapping of the field distribution around dielectrophoretically aligned cells by means of small particles as field probes. Biochim. Biophys. Acta 939:561–568.
44. Ahkong, Q. F., and Lucy, L. A. 1986. Osmotic forces in artificially induced cell fusion. Biochim. Biophys. Acta 858:206–216.
45. Lucy, J. A., and Ahkong, Q. F. 1986. An osmotic model for the fusion of biological membranes. FEBS Lett. 3548:1–11.

CHAPTER 2

Electroporation of Human Lymphoid Cells

Zdenka L. Jonak, Patrick Machy, Edward Henri,
Lynette McMillan, Steven Trulli, Deborah Matour,
and Russell Greig

I. INTRODUCTION

A. Human Monoclonal Antibodies

Routine generation of human monoclonal antibodies continues to be frustrated by the lack of antibody-producing cells and the persistent difficulties in human B-cell immortalization.[1-4] The techniques most widely employed to immortalize human B cells include classical hybridoma technology using a tumor cell fusion partner and cell transformation/immortalization accomplished by infecting the target B cells with Epstein-Barr virus (EBV).[5-7] Both methodologies have significant limitations. A suitable fusion partner for human B cells has yet to be identified, and EBV transformation is restricted to B cells that express the EBV receptor (immature B lymphocytes).[8-9]

Research at Smith, Kline & French Laboratories has focused on immortalization of human B cells by the introduction of oncogenic DNA into antibody-producing cells (oncogene technology) thus overcoming the need for a fusion partner or viral infection.[10-12] The essential steps in lymphocyte immortalization are summarized in Fig. 2-1. Immortalization of B cells depends critically on their state of activation and differentiation, the method by which DNA is transferred, and the source of oncogenic DNA/oncogene(s). The feasibility of this approach has already been demonstrated,[10-12] and the generated transfectants display several desirable features, including karyotypic stability, viral-free cell lines, and significant levels of immunoglobulin production. However, development of this technology is still in an early phase and several technical parameters have yet to be optimized. Prime among these is the DNA source for efficient and reproducible B-cell transformation/immortalization and the availability of human B lymphocytes with the desired antigenic specificity.

The emergence of electroporation has aided considerably in transfecting DNA into lymphoid cells and has expedited progress in this area significantly. This technology has been extended to incorporate a degree of cell selectivity by establishing conditions for transfecting, via electroporation, particular genes into specific lymphoid cells using targeted DNA-containing liposomes. This chapter describes the methods and conditions for electroporation-mediated DNA transfection and addresses the important experimental parameters that influence the success of this technology.

B. Gene Transfection into Mammalian Cells

Transfection of DNA into mammalian cells has contributed significantly to our understanding of gene expression and regulation. Two procedures, "natural" and "artificial," have been used to introduce exogenous genetic material into target cells. Natural transfection can be accomplished by infecting cells with genetically manipulated but intact viruses. Artificial procedures involve temporary physical or chemical perturbations of the target cell plasma membrane to permit entry of exogenous DNA. These include direct microinjection, vehicle-mediated transfer (e.g., liposomes,[13-14] erythrocyte ghosts,[15] reconstituted Sendai virus envelopes[16]), transfer aided by facilitators (e.g., calcium phosphate,[17-18] calcium phosphate/polyethylene glycol-dimethyl sulfoxide [DMSO],[10,19] DEAE-dextran[20-21]) laser microbeam,[22] and electroporation.[23-28] Each procedure is distinguished by its own limitations, and selection of the most appropriate technique depends on the biological system under manipulation. Electroporation is the method of choice for the transfection of DNA into human lymphocytes or human lymphoid cell lines.

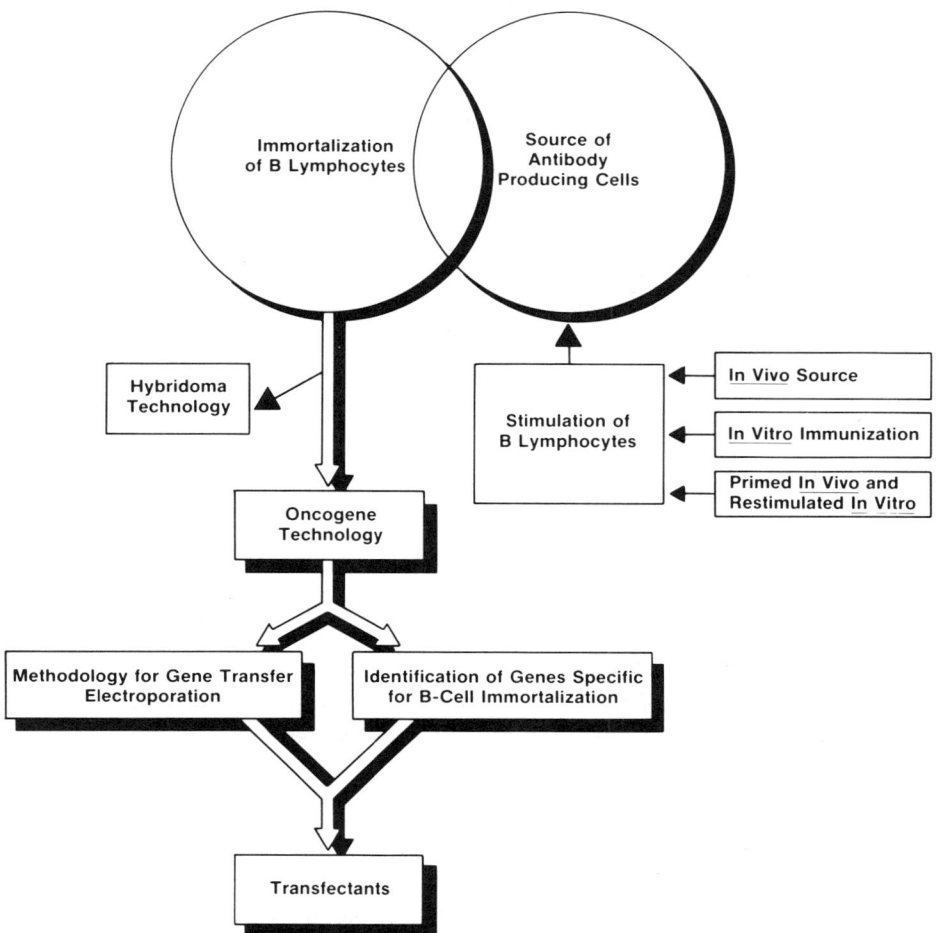

Fig. 2-1. Essential steps in lymphocyte immortalization.

II. MATERIALS/METHODS

A. Solutions/Medium

1. HY Medium[29]:

   To 500 ml of DMEM-HG (Dulbecco's Modified Eagle—high glucose) medium with 10% fetal calf serum (FCS) add:

   5 ml of L-glutamine (100×) [200mM]

   5 ml of oxaloacetate, pyruvate, insulin (OPI) (100×)
   (to prepare 100 ml of 100× OPI combine:
   CIS-oxalic acid, 1500 mg, Sigma D-7753; insulin, 2000 U, SIGMA I-6634; pyruvic acid, 500 mg, Sigma P-5280)

   5 ml hypoxanthine, thymidine (HT) (100×, Gibco 32D-1067
   (to prepare 100 ml of 100× HT, combine:
   hypoxanthine 10^{-2}M, thymidine 1.6×10^{-3}M)

   0.5 ml gentamicin (10 mg/ml, Gibco #600-5710)
   (HY-medium is also available from Gibco—#84-5123)

2. TY Medium[19]:

To 500 ml of HY medium with 15% FCS add:

0.5 ml ITS [insulin, transferrin, selenium (CR-ITS Premix, Collaborative Research)]

2 μl 2-aminoethanol (Sigma ED135)

2 μl 2-mercaptoethanol (Sigma M7522)

3. DMEM-HG medium with 15% FCS (Hyclone)
4. Sterile Dulbecco's phosphate buffered saline (PBS) pH 7.3 (Gibco #310-490)
5. 0.3 M glucose pH 7.0
6. 0.4% trypan blue (Gibco)

B. Cell Lines

The human Burkitt's lymphoma cell line, BJAB,[30] was cultured in TY medium supplemented with 15% FCS at 37°C in a 7% CO_2 atmosphere.

METHOD I: Preparation of Human Lymphocytes

Peripheral blood lymphocytes (PBLs) were the major source of human antibody producing B cells. PBLs were either isolated from healthy donors or obtained from the Red Cross (outdated blood, 24 hr old). Later experiments employed lymphocytes from tonsils and spleen cells, because large quantities can be readily obtained in one procedure. These cells were separated as described,[31] aliquoted, and frozen in liquid N_2 (95% FCS with 5% DMSO was used as the freezing mix).

Procedure for Cell Separation

1. Admix equal volume of whole blood with PBS (pH 7.2–7.4).
2. Dispense 30 ml aliquots into 50 ml centrifuge tubes.
3. Gently and slowly add 10 ml lymphocyte separation medium (LSM, Organon Teknika Corporation, #36427/8410-01) by inserting pipette into bottom of tube and creating an upper layer of blood above the LSM. Withdraw pipette and cap the tube.
4. Centrifuge for 30 min at 500 × g in horizontal centrifuge at 20°C.
5. Using Pasteur pipette, siphon out clear layer of lymphocytes.
6. Wash lymphocytes once with equal volume of PBS or TY medium. Resuspend as needed, or proceed for further separation into B, T, and accessory cells.[31]

Points to Consider

Wear gloves when handling human blood.

Process under sterile conditions.

METHOD II: In Vitro Stimulation/Activation of Human Lymphocytes

Successful production of human monoclonal antibodies depends on the availability of specifically stimulated antibody-secreting lymphocytes. These can be obtained only from patients who have been exposed to the appropriate antigen either through disease (e.g., neoplastic, bacterial/viral/fungal infections) or environmental contacts (chemicals/vaccinations).[4] In order to generate antibodies against preselected antigen(s) it is necessary to present antigen to the lymphocytes in culture, a process called *in vitro immunization*. Procedures for performing in vitro immunizations with human lymphocytes have recently been reviewed, and in these studies immortalization of stimulated lymphocytes was done using conventional hybridoma techniques (fusion with tumor partner) or by EBV infection.[5-7] An important feature of these methodologies is that the lymphocytes must be stimulated/activated prior to fusion/infection, which is also true for gene transfection.[10-12]

Because in vitro immunization is a complex system with many variables, the authors, when developing the oncogene transfection technique focused first on the generation of the transfectants, regardless of their phenotype. Thus, although the authors tested for monoclonal antibody production, the specificity of these reagents was not addressed. This chapter describes the experimental conditions for in vitro mitogen-activation of human lymphocytes and the oncogene-mediated generation of transfectants.

Procedure

Peripheral blood lymphocytes, spleen cells, or lymphocytes from tonsils were used for in vitro activation. In vitro culture conditions were as follows:

1. Human lymphocytes suspended at 10^7 cells/ml in TY medium.
2. TY medium was supplemented with 10% FCS and 10 μg/ml of lipopolysaccharide (LPS) (Difco 055:B5), concanavalin A (1 μg/ml) (ICM #79-003), or pokeweed mitogen (Gibco #670-5360).
3. Cultures were maintained in plastic T75-cm^2 flasks for 2 days at 37°C in a 7% CO_2 atmosphere.
4. Stimulated cells were tested for viability and used for electroporation as described in "Procedure" (Method V).

Points to Consider

1. Different in vitro systems can be used for the stimulation/activation of human lymphocytes, but which of these are compatible with oncogene immortalization has yet to be established. This experimental system, through the purposeful depletion of monocytes and addition of condition medium, favors the selection of antibody-producing cells.[11,32]
2. Ensure that lymphocytes are viable and stimulated or activated prior to electroporation. This is necessary for successful transfection.

METHOD III: Purification of Genomic DNA

This method is a modification of standard phenol extraction and ethanol precipitation,[19,33-34] and provides high molecular weight genomic DNA suitable for transfection and Southern blotting.

Solutions

1. Lysis solution: 1% NP40 (Sigma N-6507), 0.1 M Tris, 0.05 M EDTA, 0.1 M NaCl, pH 7.8-7.9.
2. TE buffer (100 × stock): 121.1 g Tris-base, 29.2 g EDTA, adjusted to pH 8.0 with concentrated HCl.
3. Proteinase K (BRL 5530UA): Dissolve 10 mg/ml in TE buffer (store −70°C in small aliquots).

4. Ribonuclease A (Sigma R-5125): Dissolve 10 mg/ml in distilled water. Heat treat at 80°C for 30 min (store −70°C in small aliquots).
5. Phenol (BRL #5509UA): Hydrate before use.
6. Chloroform: isoamyl alcohol are mixed in a ratio of 1:24.
7. 10% SDS in distilled water (w/v).
8. 7.5 M ammonium acetate.
9. Absolute ethanol.

Procedure

1. Wash cells (1×10^8) 3 times in ice-cold PBS without Ca^{++} and Mg^{++}. Flick cells to loosen pellet.
2. Add 4.75 ml of lysis solution plus 0.25 ml of 10% SDS. Resuspend cell pellet.
3. Add 50 μl of proteinase K (10 mg/ml) to a final concentration of 100 μg/ml.
4. Incubate at 37°C (2 hr to overnight) until solution is clear. If, after incubation, DNA solution is too viscous, add equal volume of TE buffer.
5. Add equal volume of phenol:chloroform/isoamyl alcohol (1:1) and invert several times to mix the two layers.
6. Spin down for 15 min at 2000 × g.
7. Remove the top aqueous layer to a clean tube and discard lower layer into phenol waste.
8. Repeat steps 5–7.
9. Extract with equal volume of chloroform/isoamyl alcohol. Invert several times.
10. Spin down for 15 min at 2000 × g.
11. Remove aqueous layer to a clear tube and discard lower layer.
12. Repeat steps 9–11.
13. Ethanol precipitation: add 0.5 volume of 7.5 M ammonium acetate and 2 volumes of absolute ethanol. Shake gently, incubate at room temperature for 30 min or overnight at 4°C, and spin down at 2000 × g for 15 min. Wash once with absolute ethanol.
14. Redissolve DNA in TE buffer (5 ml). Dialyze overnight in TE buffer.
15. Add 50 μl of stock RNase (10 mg/ml, previously inactivated at 80° for 30 min.). Incubate for 1 hr at 37°C.
16. Add 50 μl of stock proteinase K.
16. Incubate for 2 hr at 37°C or overnight.
17. Repeat steps 5–14.
18. Measure optical density of DNA at A_{260} and A_{280} nm. The 260/280 ratio should be between 1.8 and 2.0; 1 O.D. unit at A260 equals 50 μg of DNA.[33-34]

Points to Consider

1. Quantitation of DNA is only semiquantitative. Alternative methods exist, but none is without drawbacks. To evaluate purity and estimate DNA concentration, the absorbance is measured at two wavelengths, 260 nm and 280 nm. The ratio of A_{260}/A_{280} of a pure DNA preparation should be between 1.8 and 1.9.[33-34]
2. Overestimation of the DNA content due to contamination with protein/phenol (lower ratio), RNA (higher ratio), and other substances that absorb at 260 nm is the major drawback of this technique.
3. Resuspending DNA in aqueous solution after ethanol precipitation is difficult, and overnight incubation at 37°C is usually required to assure total DNA hydration.
4. Safety considerations demand utmost care when handling organic solvents (i.e., fume hood, rubber gloves, safety glasses, proper disposal).

C. Vectors and Selection for Transformants

The plasmid(s)[35] are eukaryotic expression vectors in which the coding regions of different genes have been inserted under control of the SV40 promoter. These vectors also carry dominant selective markers, usually the bacterial gene XGPRT (xanthine-guanine phosphoribosyltransferase) or the NEO marker genes.

1. XGPRT Marker Gene/Mycophenolic Acid Selection

In contrast to untransfected mammalian cells, which do not efficiently use xanthine for purine nucleotide synthesis, cells that are capable of expressing the transfected XGPRT gene can synthesize GMP (guanine monophosphate) from xanthine via XMP (xanthine monophosphate). After transfection, surviving cells producing XGPRT can be selectively grown with xanthine as the sole precursor for guanine nucleotide formation in a medium containing mycophenolic acid, which blocks de novo purine nucleotide synthesis.[36]

The selection medium used was TY medium (see page 34) containing 250 μg/ml xanthine (Sigma), 25 μg/ml adenine (Sigma), and 5 μg/ml mycophenolic acid (Sigma).

2. NEO Marker Gene/G418 Selection

The mechanism of this selection is the sensitivity to the G418 antibiotic (an aminoglycoside). Only cells expressing the NEO marker gene [coding for bacterial phosphotransferase enzyme APH (3′) II] will display resistance to G418, thus permitting facile selection of transfectants.[37]

The *selection medium* was TY medium (see page 34) containing 800 μg/ml G418 (Geneticin, G418 sulphate, Gibco 860-1811).

3. Points to Consider

Since mammalian cells display wide variations in their sensitivity to mycophenolic acid or G418, the concentration of drug required for selection must be individually tailored for each cell line. A simple method for achieving this is to grow the cells (in a multi-well plate) in a range of drug concentrations. The desirable level of drug is the lowest concentration that kills all the cells within 7–10 days. Cells that divide more rapidly are typically killed more readily, and this should be considered when testing for the appropriate drug concentrations.

METHOD IV: *Preparation of Targeted Liposomes Containing DNA*

This procedure involves encapsulating DNA into liposome, coating the liposomes with modified protein A, and incubating liposome-protein A complex with antibody-coated cells.

Monoclonal Antibodies (mABs)

Murine mABs B1.23.2 and H100-5/28 are of the IgG$_{2a}$ subclass. B1.23.2 is directed against nonpolymorphic determinants expressed on human major histocompatibility complex (MHC)-encoded molecules HLA-B and -C.[38] H100-5/28 has affinity for H-2K murine MHC-encoded antigens.[39] Antibodies were purified from culture supernatants by chromatography on protein A-Sepharose 4B (Pharmacia) columns.

Lipids for Preparation of Liposomes

Dimyristoyl phosphatidylcholine (DMPC, Avanti Polar Lipids), dimyristoyl phospha-tidylserine (DMPS, Avarti Polar Lipids), cholesterol (Sigma), and dipalmitoyl phospha-tidyl-ethanolamine (DPPE, Sigma) modified with N-hydroxysuccinimidyl-3-(2-pyridyl-dithio) propionate (SPDP, Pharmacia) (DPPE-DTP)[40-41] were used at the molar ratio of DMPC:DMPS: cholesterol:DPPE-DTP=54:10:35:1.

Procedure for Liposome Preparation

Large unilamellar liposomes (LUV) were prepared by reverse-phase evaporation as previously described.[42-43]

1. 40 μM of total dried lipids were dissolved in 1 ml of chloroform (Fisher Scientific) and 2 ml of isopropyl ether (Fisher Scientific).
2. 1 ml of 1.7 to 2 mg/ml of plasmid (linearized by restriction enzyme) was injected into the organic phase, and the mixture emulsified in Luer-lock syringe.
3. In some experiments an aliquot of ^{32}P-labeled plasmid or 1 ml of 20 mM purified carboxyfluorescein was used as a control.[44]
4. This mixture was reverse-phase evaporated at 40°C and 400 mm Hg under a stream of N_2.
5. The liposome preparation was filtered under N_2 pressure through a 0.4 μm poly-carbonate membrane (Nucleopore) using a stirred ultrafiltration cell (Amicon).
6. Liposomes were mixed with SPDP-modified protein A (Pharmacia) (10 moles SPDP/mole protein A); (200 μg protein A plus an aliquot of ^{125}I-labeled protein A) as described previously.[40-41,43,45] The liposome preparations were dialyzed overnight in 145 mM NaCl, 10 mM HEPES buffer, pH 8.0.
7. Liposomes were then incubated with 50 μg DNase I (Sigma) and 10 mM $MgCl_2$ for 30 min at room temperature before chromatography on Sepharose 4B columns to remove uncoupled and unencapsulated molecules.
8. Liposomes were sterilized by filtration through 0.44 μm Millipore filters and stored at 4°C.

Points to Consider

Between 5-8% of DNA and carboxyfluorescein was encapsulated, corresponding to about 1–3 molecules of plasmid per liposome, and 2–5 protein A molecules coupled per liposome.

D. Transfection Protocols

METHOD V: Electroporation

Electroporation is an effective method for introducing genes into human lympho-cytes/human lymphoid cell lines. Several electroporation systems and conditions have been described.[11-12,23-28] Here we document our standard procedure.

Equipment

- BTX Electro Cell Manipulator, model 401-AM (BTX, Biotechnologies—Experimental Research, Inc., 3742 Lewell Street, San Diego, California, U.S.A.)
- Chambers for electroporation:
    Coaxial chamber 420 (1 ml volume)
    Flat chamber 481 (0.5 ml volume/1.0 mw gap)

- BTX Optimizor 500 (graphic pulse analyser for monitoring electroporation)
- LKB 2197 power supply and disposable plastic cuvettes and electrode as described by Potter.[24,28]

Electroporation Conditions

The BTX Electro Cell Manipulator uses a DC electric pulse to create membranous pores through which passage of DNA can occur. Cell viability and transfection efficiency depend on pulse number, amplitude, and duration; this relationship must be established for each cell line. For B cells suspended in PBS, optimal conditions were an amplitude of 500–999 (2.4 kV/cm–4.2 kV/cm) and a pulse width of 99 μsec. A number of consecutive pulses were delivered. High-intensity or prolonged pulses decrease cell viability. The suspending medium also affects viability. At an amplitude of 999 and pulse width of 99 μsec, viability of human BJAB cells was 85% in PBS and 13% in 0.3 M glucose.

Procedure

1. *Preparation of Cells:*
   a. Prepare human lymphocytes as described previously (Method I). Stimulate cells in vitro with antigen/mitogen.
   b. Grow cells in HY medium with 15% FCS. Split cells and refeed day before electroporation.
2. Wash cells 2× with ice-cold PBS and resuspend cells in ice-cold PBS at 10^7 cells/ml (ensure 95% cell viability).
3. Add 5–10 μg of plasmid DNA (supercoiled or linearized plasmid) per 5×10^6 cells in 0.5 ml chamber. Alternatively, use 10–50 μg of genomic DNA (sheared through a 26-gauge needle 3 times). *Note:* Prior to use, DNA was ethanol precipitated and resuspended in sterile PBS. Use 10–50 μg of DNA per 5×10^6 cells in 0.5 ml chamber. Mix well and keep on ice for 10 min.
4. Rinse chamber (previously stored in 70% ethanol) thoroughly with cold PBS. Place chamber on ice and connect to electrocell manipulator.
5. Add cell-DNA mixture to chamber (0.5 ml) and deliver single/multiple pulse(s). (Under these conditions approximately 75% cells are viable after 1 hr at 37°C.)
6. Transfer electroporated cells into tube and keep on ice for 10 min.
7. Add required amount of TY medium (see model) and transfer cells into culture dishes.
8. After 24 hr add appropriate selection medium (mycophenolic acid for XGPRT or G418 for NEO selection). Alternatively, use TY medium if genomic DNA without marker gene was used.
9. Replenish medium weekly. If selection is used, clones should appear within 3 weeks. For B-cell transfectants with genomic DNA allow up to 8 weeks.

Point to Consider

Important parameters and optimal conditions are described under "Model System" (Section IV).

METHOD VI: Electroporation with Targeted Liposomes

Large unilamellar liposomes, coated with protein A and encapsulating the marker gene, were used in combination with electroporation to specifically deliver DNA into target cells.[46] The procedures for binding the liposomes to target cells and electroporation conditions used for gene transfection are described below.

Binding of Targeted Liposomes to Target Cells

1. Cells (5×10^6) were incubated with 20 μg anti-HLA or anti-H-2K (control) monoclonal antibodies or incubated without antibody.
2. After 1 hr incubation at 4°C, cells were washed twice with culture medium and incubated with different concentrations of protein A-bearing liposomes containing plasmid DNA or carboxyfluorescein. (In control experiments, 50 μg of free protein A was added before incubation with liposomes.)
3. After 1 hr incubation at 4°C, cells were washed 3 times with PBS, and lysed with 1% Triton X-100. Samples were centrifuged and liposome binding to the cells was evaluated by measuring the fluorescence of the carboxyfluorescein released into the supernatants with a Perkin Elmer 650 10S spectrofluorimeter (excitation wavelength, 488 nm; emission wavelength, 520 nm.

Transfection Protocol

1. Cells were washed with PBS (Ca^{2+} and Mg^{2+} free) and incubated for 1 hr at 4°C in medium only (control)/or with 20 μg of antibodies (anti-HLA or control anti-H-2K).
2. Cells were washed with PBS and incubated in PBS for 1 hr on ice with different concentrations of protein A-bearing liposomes containing linearized DNA and/or carboxyfluorescein (control).
3. The cells were untreated (control) or electroporated using an LKB 2197 power supply (in early experiments). The voltage was adjusted to 2500 V and the current maintained at 0.8 mA for approximately 100 μsec. (Electroporation can be done in BTX electrocell manipulation under the same conditions as described for BTX.)
4. Five successive pulses were delivered to each cell preparation (LKB power supply).
5. The cells were incubated on ice for an additional 10 min and distributed into 96-well Costar microtiter plates at a density of 5×10^4 to 10^5 cells per well in TY medium.
6. After a 24-hr incubation at 37°C, cells were cultured in TY selection medium (Section II-F) supplemented with 10% FCS. The medium was replenished every 2 days, and after 15 days in culture the growth-positive wells were counted.

III. MODEL SYSTEM FOR ELECTROPORATION/ ELECTROPORATION WITH TARGETED LIPOSOMES

For our model system we selected a cell line, the human Burkitt's lymphoma, BJAB, that closely mimics our eventual cell target, namely the human B-lymphocyte.[11-12,46] BJAB cultures have been phenotyped as an early B cell,[12] and thus the authors considered it important to identify the electroporation conditions in this model system before applying them to the more demanding task of immortalizing human B-lymphocytes (see Tables 2-1 and 2-2).

To provide an objective measure of transfection efficiency, the authors selected a plasmid that carried the marker XGPRT gene. Because only cells transfected with the XGPRT gene proliferate in the selection medium, the number of proliferating colonies provides a direct measurement of transfection efficiency.

Table 2-1. Model System for Electroporation

Objectives	Example
1. Use cell lines most closely mimicking phenotype of target cell	BJAB (human Burkitt's lymphoma, B-cell type)
2. Choose appropriate selectable plasmid (dominant selection)	XGPRT gene marker
3. Prior to electroporation, determine the concentration of the drug used for selection	$5 \mu g/ml$ of mycophenolic acid (Section II-F)
4. Rapid proliferation/stimulation of cells (close to 100% viability) is necessary for transfection	Split cells day before transfection and refeed with fresh medium
5. Explore optimal electroporation conditions: medium amplitude pulse width pulse number quantity of DNA cell number plated per well	Best parameters for electroporation: PBS (viability) 500 (2.4 kV/cm) 99 μsec 2 pulses $10 \mu g/5 \times 10^6$ cells 1×10^5 per well (96–well plate)
6. Refeed cells	Refeed weekly
7. First transfectants generated	Clones appear between 10–21 days after transfection
8. Isolate/scale-up clones, test for presence/expression of marker gene/gene product	Use Southern blot test for XGPRT gene
	Use ELISA for mAB detection

To establish optimal electroporation conditions, several parameters need to be examined:

1. *Cell viability in the electroporation medium/buffer.* It was important to maintain viability of all cell types, since the human B cells are only a small portion of total PBLs. We could not establish the electroporation condition for primary B-lymphocytes, and therefore BJAB cells were used as a model system. Our results suggest that PBS was the optimum medium.

2. *The effect of cell and DNA concentrations on transfection efficiency (PBS was used as an electroporation buffer).* To estimate the appropriate cell/DNA concentrations, the authors used 5×10^6 cells in 0.5 ml chamber and varied the concentration of plasmid DNA from 1 to 50 μg. Optimal concentration was about 10 μg.

3. *The effect of electric field strength (amplitude) and duration of pulse (pulse width).* In 0.5 ml chamber; 5×10^6 cells with 10 μg of plasmid DNA in PBS buffer (pulse width of 99 μsec) were subjected to different electric field strengths (variation of amplitude) from 250–999 (1.5 KV–4.2 KV/cm). Optimal conditions were between 500 and 750 (2.4 KV–3.2 KV/cm).

4. *The number of pulses delivered.* The optimal conditions: 5×10^6 cells/PBS buffer/10 μg of plasmid DNA/500 amplitude (2.4 KV/cm)/99 μsec pulse width

Table 2-2. Model System for Electroporation with Targeted Liposomes
Objectives/Results

Objectives	Example
1. Use cell lines most closely mimicking phenotype of target cell	BJAB (human Burkitt's lymphoma, B-cell type)
2. Test the parameters and follow objectives as described in "Model System" for electroporation	Targeted liposome approach was evaluated in the same system as described for electroporation.[46]
3. Choose appropriately selective reagents (mABs, protein A) to target DNA-liposomes to defined cell type	Liposomes were coated with protein A (selective recognition of mouse IgG_{2a} antibodies), and the BJAB cells with murine mAB B1.23.2 (directed against human HLA-B and -C determinants).
4. Use plasmid DNA or small DNA fragments for liposome encapsulation	DSPK plasmid (6.3 Kb) carrying XGPRT marker gene was linearized by Sma I restriction enzyme.
5. Use appropriate controls to assure specificity/quantity of the DNA per liposome/number of targeted cells; e.g., nonspecific mABs, ^{32}P-labeled plasmids, and carboxyfluorescein	Murine mAB (H100-5/28) of same class (IgG_{2a}) was used to control nonspecific binding; ^{32}P-labeled plasmid and carboxyfluorescein measured the amount of DNA/carboxyfluorescein encapsulated and the number of liposomes bound per cell.
6. Electroporate/select clones/test for specificity	Clones of transfected cells grow in selection medium. Southern blot test proved presence of XGPRT gene.

in 0.5 ml chamber were used in experiments in which pulse number varied. Two pulses delivered consecutively were optimal.

Points to Consider

Many additional parameters play an important role in establishing optimal conditions; for example, the number of electroporated cells plated per well (10^5/well is our optimal concentration); cell viability; degree of stimulation/activation; cell type/cell cycle stage; cell manipulation during electroporation (prolonged incubation of cells in PBS, rapid pipetting); use of linearized versus supercoiled plasmid DNA (higher transfection efficiency is currently associated with linearized plasmid DNA).[24,28,46]

IV. CONCLUSIONS

A. Electroporation Conditions and Various Cell Types

During electroporation, cell membranes are subjected to a high-voltage electric field that results in a temporary loss of membrane integrity and the formation of pores, which allow exogenous DNA to enter the cell. Pore closure is a natural decay process that can be delayed by incubating the cells at 0°C. After the DNA enters the cell, a proportion becomes integrated into the host genome and generates a permanently transfected cell line.

Cells of fibroblastic origin readily take up DNA using traditional transfection methodologies.[11,28] The authors compared the efficiency of calcium phosphate coprecipitation and electroporation in transfecting a hamster fibroblast cell line (R1610). Although R1610 cells incorporate genetic material without additional manipulation,[11] both electroporation and calcium phosphate precipitation cause an equivalent increase in transfection efficiency. However, electroporation was more efficient than calcium-phosphate precipitation in transfecting Chinese hamster ovary cells (CHO); this finding indicates that the method of DNA transfer should be tailored to individual cell types.

Cells of lymphoid origin are traditionally refractory to most gene transfer procedures.[11-12,46] The present study, however, shows that electroporated BJAB cells were transfected at high efficiency with DSPK plasmid DNA. A large number (75%) of the transfectants displayed colony formation in the 96 well proliferation assay. The method was rapid and reproducible. In contrast, BJAB cells incubated with either plasmid DNA alone or calcium phosphate DNA coprecipitate failed to integrate DNA and did not form proliferating colonies.[11-12] Compared with other transfection methods[10,15-21] electroporation-mediated gene transfer delivers a low gene copy number per cell.[24,28] This was confirmed by a DNA dot blot, which demonstrated approximately one copy of the XGPRT gene per BJAB cell.

B. Electroporation with Targeted Liposomes

Electroporation is an efficient technique that can be used to transfect many cell types. However, no selectivity with regard to cell type in a heterogenous cell population can be achieved using standard electroporation conditions. It has been demonstrated that B-lymphocytes can be specifically transfected in vitro using an electric field (electroporation) and targeted liposomes containing DNA. As a model system we used large unilamellar liposomes in which plasmid DNA carrying the bacterial xanthine-guanine phosphoribosyl transferase (XGPRT) gene had been encapsulated. These liposomes were directed to target cells by surface-coupled monoclonal antibodies, and specific transfection was achieved by electroporation.[46] The use of electroporation and targeted liposomes is a simple and specific technique that provides a methodology for introducing defined biological macromolecules into specific cell types. One direct application of this technology is to specifically target the liposomes/DNA to B-lymphocytes, which represent only a small percentage of the total peripheral blood lymphocyte population, in order to capture and immortalize these cells. This approach will increase the efficiency of human monoclonal antibody production.

C. Human Transfectants/Monoclonal Antibody Production

We have generated a panel of human transfectants by stimulating mixed population of lymphocytes with mitogens and transfecting these stimulated/activated cells with oncogenic DNA/oncogene(s). The transfectants displayed various phenotypes, indicating that they were immortalized at different stages of their differentiation cascade. The karyotype of these cells appeared normal. The observations that these transfectants are also phenotypically stable and secrete immunoglobulin make them good candidates for routine monoclonal antibody production.[11-12] Before this can be achieved, however, several issues need to be resolved, including identifying an effective source of oncogenic

DNA/oncogene(s) for immortalization of B cells and the development of a more efficient in vitro immunization system. In this system, B cells respond to the desired antigens by producing antibodies of acceptable specificity, affinity, and immunoglobulin class.

In conclusion, electroporation has accelerated progress in developing suitable methods for transfecting DNA into human activated/stimulated lymphocytes. This technology has also presented additional opportunities, including the generation of new cell lines (transfectants) and their products, which aid scientists in understanding the biology of gene regulation and expression and the interaction between cells.

ACKNOWLEDGMENT

We are grateful to Jo Anne Mackey for her expert assistance in the preparation of this manuscript.

REFERENCES

1. Dorfman, N. A. 1985. The optimal technological approach to the development of human hybridomas, J. Biol. Response Modif. 4:213–239.
2. Carson, D. A., and Freimark, B. D. Human lymphocyte hybridomas and monoclonal antibodies. In Advances in Immunology. F. J. Dixon, ed., Academic Press, New York, 38:275–311, 1986.
3. O'Hare, M. J., and Yiu, C. Y. 1987. Human monoclonal antibodies are cellular and molecular probes; a review, Moll Cell. Probes 1:33–54.
4. James, K., and Bell, G. T. 1987. Human monoclonal antibody production; current status and future prospects, J. Immunol. Methods 100:5–40.
5. Kozbor, D., and Roder, J. C. 1983. The production of monoclonal antibodies from human lymphocytes, Immunol. Today 4:72–79.
6. Rosen, A., Persson, K., and Klein, G. 1983. Human monoclonal antibodies to genus-specific chlamydial antigen produced by EBV-transformed B-cells, J. Immunol. 130:2899–2902.
7. Smith, L. J., Braylan, R. C., Edmundson, K. B., Nutkis, J. E., and Wakeland, E. K. 1987. *In vitro* transformation of human B-cell follicular lymphoma cells by Epstein-Barr virus, Cancer Res. 47:2062–2066.
8. Frade, R., Barel, M., Ehlin-Henriksson, B., and Klein, G. 1985. gp140, the C3d receptor of human B lymphocytes, is also the Epstein-Barr virus receptor, Proc. Natl. Acad. Sci. USA. 82:1490.
9. Chan, M. A., Stein, L. D., Dosch, H. M., and Sigal, N. H. 1986. Heterogeneity of EBV-transformable human B lymphocyte populations, J. Immunol. 136:106.
10. Jonak, Z. L., Braman, V., and Kennett, R. H. 1984. Production of continuous mouse plasma cell line by transfection with human leukemia DNA, Hybridoma 3:107.
11. Jonak, Z. L., Owen, J. A., and Machy, P. Strategies for the immortalization of B lymphocytes. In In Vitro Immunization in Hybridoma Technology. C. A. K. Borrebaeck, ed., Elsevier/North Holland, Amsterdam, 1988.
12. Jonak, Z. L., Owen, J. A., Machy, P., Leserman, L. D., and Russell, G. 1988. Gene transfection and lymphocyte immortalization: a new approach to human monoclonal antibody production, Advanced Drug Delivery Reviews 2:207–228.
13. Poste, G., Papahadjopoulos, D., and Vail, W. J. 1976. Lipid vesicles as carriers for introducing biologically active materials into cells, Methods Cell Biol. 14:33.
14. Schaefer-Ridder, M., Wang, Y., and Hofschneider, P. H. 1982. Liposomes as gene carriers: Efficient transformation of mouse L cells by thymidine kinase gene, Science 215:166–169.
15. Furusawa, M., Yamaizumi, M., Nishimura, T., and Uchida, Y. 1976. Use of erythrocyte ghost of injection of substances into animal cells by cell fusion, Methods Cell Biol. 14:73.

16. Loyter, A., Tomasi, M., Gitman, A. G., Etinger, L., and Nussbaum, O. 1984. The use of specific antibodies to mediate fusion between Sendai virus envelopes and living cells, Ciba Foundation Symposium 103:163.

17. Graham, F. L., and Van der Eb, A. J. 1973. A new technique for the assay of infectivity of human adenovirus 5 DNA, Virology 52:456.

18. Chen, C. A., and Okayama, H. 1988. Calcium phosphate-mediated gene transfer: A highly efficient transfection system for stably transforming cells with plasmid DNA, BioTechniques 6(7):632–638.

19. Jonak, Z. L., and Kennett, R. H. Methods for transfection of human DNA into primary mouse lymphocytes and NIH/3T3 mouse fibroblasts, In Monoclonal Antibodies and Functional Cell Lines. R. H. Kennett, K. B. Bechtol, and T. J. McKearn, eds., Plenum Press, New York, 1984.

20. Gopal, T. V. 1985. Gene transfer method for transient gene expression, stable transformation, and cotransformation of suspension cell cultures, Molecular and Cellular Biology 5(5):1188–1190.

21. McCutchan, J. H., and Pagano, J. S. 1968. Enhancement of the infectivity of simian virus 40 deoxyribonucleic acid with diethyl-aminoethyl-dextran, J. Natl. Cancer Inst. 41:351–356.

22. Tsukakoshi, M., Kurata, S., Namiya, Y., Ikawa, Y., and Kasuya, T. 1984. A novel method of DNA transfection by laser microbeam cell surgery, Applied Physics B. 35:2284.

23. Neumann, E. 1984. Electric gene transfer into cultured cells, Bioelectrochemistry and Bioenergetics 13:219.

24. Potter, H., Lawrence, W., and Leder, P. 1984. Enhancer-dependent expression of human kappaimmunoglobulin genes introduced into mouse pre-B lymphocytes by electroporation, Proc. Natl. Acad. Sci. USA 81:7161.

25. Tur-Kaspa, R., Teicher, L., Levine, B. J., Skoultchi, A. I., and Shafritz, D. A. 1986. Use of electroporation to introduce biologically active foreign genes into primary rat hepatocytes, Molecular and Cellular Biology 6(2):716–718.

26. Andreason, G. L., and Evans, G. A. 1988. Introduction and expression of DNA molecules in eukaryotic cells by electroporation, BioTechniques 6(7):650–659.

27. McNally, M. A., Lebkowski, J. S., Okarma, T. B., and Lerch, L. B. 1988. Optimizing electroporation parameters for a variety of human hematopoietic cell lines, BioTechniques 6(9):882–886.

28. Potter, H. 1988. Electroporation in biology: Methods, applications, and instrumentation, Analytical Biochemistry 174:361–373.

29. Kennett, R. H. Fusion by centrifugation of cells suspended in polyethylene glycol. In Monoclonal Antibodies, Hybridomas: A New Dimension in Biological Analyses, Plenum Press, New York, 1980.

30. Klein, G., Lindahl, T., Jondal, M., Leibold, W., Menezes, J., Nilsson, K., and Sundstrom, C. 1974. Continuous lymphoid cell lines with characteristics of B cells (bone-marrow-derived), lacking the Epstein-Barr Virus genome and derived from three human lymphomas, Proc. Natl. Acad. Sci. USA 71(8):3283–3286.

31. Danielsson, L., Moller, S. A., and Borrebaeck, C. A. K. 1988. Effect of cytokines on specific *in vitro* immunization of human peripheral B lymphocytes against T-cell dependent antigens, Immunology 61:51–55.

32. Owen, J. A., Muirhead, K., Jensen, C., and Jonak, Z. L. Flow cytometric analysis of human peripheral blood cultures during *in vitro* immunization, in preparation.

33. Maniatis, T., Fritsch, E. F. and Sambrook, J. 1982. In Molecular Cloning, A Laboratory Manual, Cold Spring Harbor Laboratory, pp. 468–469.

34. Schleif, R. F., and Wensink, P. C. Measuring Nucleic Acid Concentration and Purity, In Practical Methods in Molecular Biology, Springer-Verlag, New York, 1984.

35. Pfarr, D. S., Sathe, G., and Reff, M. 1985. A highly modular cloning vector for the analysis of eukaryotic genes and gene regulating elements, DNA 4:461–467.

36. Mulligan, R. C. and Berg, P. 1981. Selection for animal cells that express the E. coli gene coding for xanthine-guanine phosphoribosyl-transferase, Proc. Natl. Acad. Sci. USA 78:2072–2076.

37. Davis, L. G., Dibner, M. D., and Battey, J. F. 1986. Selection of Transfected Mammalian Cells: The G418 Method, In Molecular Biology, Elsevier, Amsterdam, The Netherlands, 1986.

38. Rebai, N., and Malissen, B. 1983. Structural and genetic analysis of HLA Class I molecules using monoclonal Xeno antibodies, Tissue Antigens 22:107–117.

39. Lemke, H., Hammerling, G. J., and Hammerling, U. 1979. Fine specificity analysis with

monoclonal antibodies of antigens controlled by the major histocompatibility complex and by the Qal TL region in mice, Immunol. Rev. 47:175–206.

40. Leserman, L. D., Barbet, J., Kourilsky, F., and Weinstein, J. N. 1980. Targeting to cells of fluorescent liposomes covalently coupled with monoclonal antibody or protein A, Nature (London) 288:602–604.

41. Barbet, J., Machy, P., and Leserman, L. D. 1981. Monoclonal antibody covalently coupled to liposomes: specific targeting to cells, J. Supramol. Struct. Cell Biochem. 16:243–258.

42. Machy, P., and Leserman, L. D. 1983. Small liposomes are better than large liposomes for specific drug delivery *in vitro*. Biochim. Biophys. Acta 730:313–320.

43. Truneh, A., and Machy, P. 1987. Detection of very low receptor numbers on cells by flow cytometry using a sensitive staining method, Cytometry 8:562–567.

44. Ralston, E., Hjelmeland, L. M., Klausner, R. D., Weinstein, J. N., and Blumenthal, R. 1981. Phase transition release, a new approach to the interaction of proteins with lipid vesicles. Application to lipoproteins, Biochim. Biophys. Acta 649:133–137.

45. Machy, P., Pierres, M., Barbet, J., and Leserman, L. D. 1982. Drug transfer into lymphocytes mediated by liposomes bound to distinct sites on H-2 encoded I-A, I-E and K molecules, J. Immunol. 129:2098–2102.

46. Machy, P., Lewis, F., McMillan, L., and Jonak, Z. L. 1988. Gene transfer from targeted liposomes to specific lymphoid cells by electroporation, Proc. Natl. Acad. Sci. USA 85:8027–8031.

CHAPTER 3

Formation of Hybridomas Secreting Human Monoclonal Antibodies with Mouse-Human Fusion Partners

Susan Perkins, Ulrich Zimmermann, Petra Gessner, and Steven K. H. Foung

I. INTRODUCTION

The development of techniques to produce murine monoclonal antibodies and their application has led to advances in many fields of biomedical research. Although it is likely that human monoclonal antibodies (HMAbs) will have distinct advantages, particularly in therapeutic applications, the ability to produce HMAbs has lagged far behind those developed for the murine system. Three general problems have limited the development of HMAbs: the lack of a fusion partner to immortalize human B cells, stability of Ig secretion, and the rarity of antigen-specific B cells in peripheral blood and lymphoid tissues. Two approaches that have been explored extensively to immortalize B cells are transformation by Epstein-Barr virus (EBV) and fusion of human B cells with suitable murine, human, or human lymphoblastoid cell lines (reviewed in James and Bell[1]). The system we have developed to overcome these obstacles involves the use of mouse-human heteromyelomas fused to appropriately activated B cells obtained in vivo (e.g., during acute infection), or after initial expansion by in vitro activation. This strategy has been effective in the production of HMAbs to a variety of antigens using polyethylene glycol (PEG) as a fusogen (Table 3-1).

For many other antigen systems, however, this approach has not been successful, mainly because of the rarity of antigen-specific B cells. The problem is partly overcome by expanding the B-cell pool with in vitro activation. Another solution is the development of cell fusion techniques capable of achieving a high fusion efficiency with a small number of input B cells. This can be accomplished with electric field-induced cell fusion or electrofusion as a highly efficient alternative to PEG-induced fusion in hybridoma formation.

In a comparative study of fusion efficiency, EBV-activated B cells were fused to a mouse-human heteromyeloma, SBC-H20, by electrofusion or PEG (Table 3-2).[11] Using $3-6 \times 10^6$ input B cells, 3-16 hybridomas per 10^5 B cells were produced with PEG. For electrofusion, the required number of input B cells was substantially lower ($1-3 \times 10^6$) with higher fusion efficiency of 30 to more than 200 hybridomas per 10^5 B cells. A similar set of experiments using pokeweed mitogen (PWM)-stimulated B cells produced similar results. PEG fusions produced 1-10 hybridomas per 10^5 B cells. The yield by electrofusion was 15-53 hybridomas per 10^5 B cells. The actual fusion efficiency varied with the specific cells and method of stimulation. The implementation of this technique has led to the successful production of a panel of HMAbs to human cytomegalovirus (HCMV), using two different mouse-human heteromyelomas.[10]

The principles and potential fields of application of electrofusion have been extensively reviewed[12,13]; this chapter presents the experimental conditions that, if carefully followed, will produce a high hybrid yield with activated human B cells and mouse-human fusion partners by electrofusion. The authors have had the most experience using two mouse-human partners: SBC-H20 and $K_6H_6/B5$.[9] The fusion procedures described have been optimized for these cell lines. Examples of fusion parameters throughout the chapter will be those developed for the production of HMAbs to HCMV.

The Zimmermann Cell Fusion System (see equipment section) consists of a power supply that allows researchers to vary the AC alignment and DC fusion currents and different fusion chambers with electrodes 200 μm apart. Since the distance that the current must travel is constant in all the fusion procedures described, voltage will be described as volts (as indicated on the power supply) and not as volts or kilovolts/centimeter. Voltages that will cause the cells to fuse depend on the cell lines, the medium in which they are fused, cell number, size, state of activation, stage in the growth cycle, and so on. An advantage of this system is that the voltages applied can be varied depending on the cell populations fused. The two parameters that have proven most useful to vary have been the DC voltage and the alignment time.

Table 3-1. Human Monoclonal Antibodies Derived from Mouse-Human Heteromyelomas

Fusogen	Antigen	Ig Class	Reference
PEG	Type A red blood cells	IgM	2
	$Rh_o(D)$	IgG, IgM	3
	rh[G]	IgG	4
	T cell	IgM	5
	VZV	IgG	6
	EBV	IgM	7
	Mycobacterium leprae	IgM	7
	HIV	IgG	8
	B cell tumors	IgM	9
Electrofusion	VZV	IgG	*
	HCMV	IgG	10

*S. Perkins and S. Foung, manuscript in preparation.

Fusions are performed using two different sets of media (see section on media): (1) Zimmermann fusion medium (ZFM) consisting of a sugar-containing phosphate buffer of low conductivity for fusion, followed by a postfusion medium used to aid in cell recovery, and (2) L3 fusion medium, an isotonic sugar-containing buffer with no postfusion recovery medium.

Fusion steps consist of counting cells, washing them in fusion medium, fusion, postfusion incubation (with or without a change in medium), harvesting the cells from the chamber, and plating. This is a long and traumatic process for the cells. Particular attention to small details is required to minimize cell loss.

1. Have all equipment and reagents ready and available for use. It is important to minimize the length of time the cells spend in suboptimal growing conditions, such as at room temperature without CO_2, in cell pellets, or in washing or fusion media without nutrients.
2. Since many cells will adhere to glass, use plastic serologic pipettes or tips to work with the cells.

Table 3-2. Comparative Efficiency of PEG and Electric Field-induced Cell Fusions

Activation	Fusogen	Input B cells[a] (10^6)	Wells with hybrids/10^6 B Cells	Interquartile range of colonies[b]	Fusion efficiency[c]
EBV	PEG	3.0–6.0	26–54	1–3	3–16
	Electrofusion	1.0–3.0	63–229	5–10	32–229
PWM	PEG	1–2.3	12–35	1–3	1–10
	Electrofusion	1.0	76–105	2–5	15–53

[a]The range in number of B cells used in fusion.
[b]An estimated range of observed colonies per well in over 50% of the wells.
[c]The range in number of hybrids per 10^5 input B cells.

3. Resuspend cell pellets immediately after washes, pipetting gently and avoiding air bubbles.
4. Know the centrifugation force necessary to pellet your cells optimally in fusion media with your equipment to reduce cell loss in the wash steps. Cells must be washed free of serum protein prior to fusion.
5. Adhere to the time limits for cells in fusion media as described in the methods section. Cell viability will drop with longer exposure times.

Because the B cells to be fused are limited in number and the exact yield cannot be predicted ahead of time, rigid rules as to exact cell number, fusion ratio, and plating density for each fusion are not made. Fusions are performed using more fusion partner cells than B cells. When fusing very few B cells, increase the number of fusion partner cells to ensure that the mixture of cells will pellet. Wash volumes vary, depending on the cell number being fused. Plating densities are determined after an actual cell count is made, and range from 2-9×10^4 cells/microtiter well. Lower densities are used to plate fusions that generally give greater hybrid yield, such as those with EBV-activated B cells. The higher densities are used with more stringent fusion conditions, such as those used to produce hybrids from B cells activated with HCMV antigen and PWM and fused at 70-90 DC volts.

Equipment (including the power supply and its settings), media composition, and cell preparation for preparing fusion procedures are described first. Fusion Methods I and II in the helical chambers are the procedures that were used to generate the anti-HCMV hybrids. Method III is a procedure for fusing very small numbers of cells (10^6 cells or less). Method IV is a procedure for using the open fusion chamber to derive optimal fusion parameters for different cells.

II. EQUIPMENT

A. Electrofusion Systems

The Z1000 Zimmermann Cell Fusion System was used to develop this methodology in our laboratory and was purchased from GCA/Precision Scientific Group (Chicago, IL).[14] Similar equipment can be obtained from B. Braun Biotech (Allentown, PA).

1. Power Supply

The Zimmermann Z1000 power supply allows one to manipulate the electrical currents used for cell fusion in a variety of ways:

Duty cycle: 1-100%

Alignment voltage limit: 0-50

Preset or continuous alignment

Alignment frequency: 10 kHz-5 MHz

Alignment voltage: 0-35 V

DC fusion voltage: 0-250 V

Fusion pulse duration: 0-99.9 μs

Alignment off time between fusion pulses: 0-999 ms

Number of fusion pulses: 1-9

Time between fusion pulses: 0.1-9.9 s

For fusion with mouse-human heteromyelomas, the following parameters have been used successfully:

Duty cycle: 100%

Alignment voltage limit: 50

Preset alignment: 20–30 s

Alignment frequency: 1000 kHz

Alignment voltage: 5 V

DC fusion voltage: 30–90 V

Fusion pulse duration: 15.0 μs

Alignment off time between fusion pulses: 10 ms

Number of fusion pulses: 3

Time between fusion pulses: 1.0 s

All parameters are preset except the alignment voltage adjust, which is quickly turned from 0 to 5 V immediately after the alignment switch is turned on. Preset alignment causes gradual tapering off of the alignment current after the DC fusion pulse. The power supply should be turned on prior to cell preparation for fusion.

2. Fusion Chambers

    a. Open chamber (Fig. 3-1)

        (1) Polystyrene slide containing 2 parallel electrodes of 200 μm diameter, 200 μm apart, and 200 μm above the slide surface.

        (2) Cover slips—18 mm² (Corning)

        (3) To clean: Chambers are reusable, but cannot be autoclaved.

            (a) Immediately after use, soak in Alconox (VWR) and water solution for several hours (the solution should contain enough Alconox to make it feel slightly slippery).

            (b) Rinse with water.

            (c) Rinse with double distilled water.

            (d) Soak in double distilled water for several hours.

            (e) Air dry.

            (f) Rinse with 95% ethanol just prior to use and air dry.

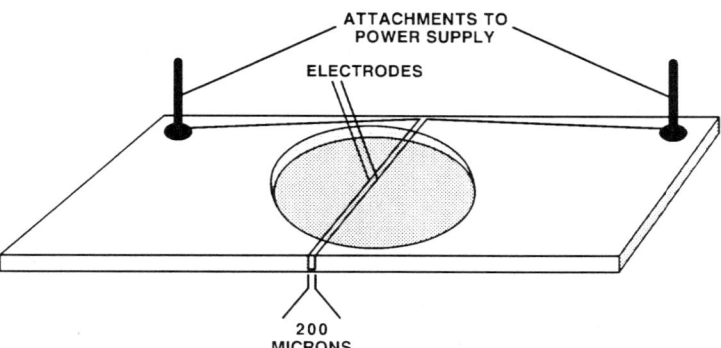

Fig. 3-1. Open fusion chamber.

a. Helical chambers (Fig. 3-2)

   (1) 200-μm helical chamber

      The electrode assembly consists of a 200-μm diameter platinum electrode wound in a helix with 200-μm spacing and two protruding electrodes that can be attached to the power supply. The outer receptacle holds the cells.

   (2) 200-μm custom micro helical chamber

      This chamber has the same 200-μm helical electrode assembly with a slightly smaller outer receptacle and less dead space.

   (3) To clean:

      (a) Immediately after use, place in Alconox and water solution. Soak in this solution overnight.

      (b) If the electrodes have large deposits on them, rub gently with fine emery cloth.

      (c) Rinse with distilled water.

      (d) Soak in double distilled water over a second night.

      (e) Air dry.

      Note: This soaking method does not clean everything out of the helix, and eventually debris will start to collect. A sonic water bath will remove all the debris, but repeated use will also cause the helix to unwind. Compromise: 30–60-s exposure in a sonic water bath after the overnight soak in Alconox is used only occasionally.

   (4) Sterilization

      (a) Autoclaving—wrap chamber parts separately in autoclave paper and place in a vented container (a tube that will withstand autoclaving covered with autoclave paper works well). Autoclave at 121°C and 15 psi on fast exhaust for 15 min.

      (b) Alcohol—Soak in 95% ethanol for 10 min and air dry for 1 hr under the sterile hood.

      Note: Alcohol will not kill all bacterial spores.

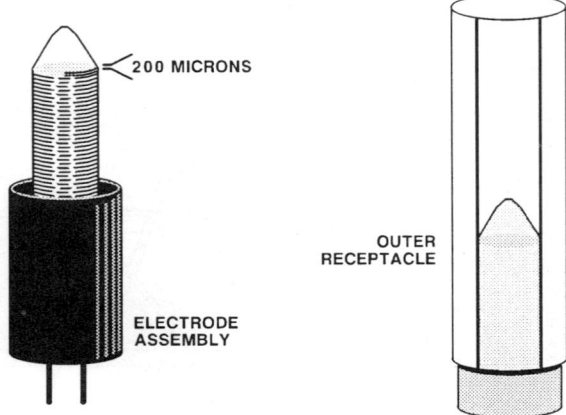

Fig. 3-2. Helical fusion chamber.

B. Miscellaneous Equipment

Timing is critical, so it is important to have all equipment available and ready for use. Label as many tubes and plates as possible before starting to work with the cells. Equipment needed includes:

1. Sterile hood to use for performing all cell manipulations.
2. Incubator for cells: 37°C; 6-6½% CO_2.
3. Microscope for counting cells, observing cells in culture, and working with the open chamber.
4. Centrifuges to use at room temperature to pellet cells for counting and for fusions with L3 fusion medium; set at 10°C for fusion medium washes with ZFM.
5. 37°C water bath to prewarm Zimmermann postfusion (ZPFM) and L3 fusion media.
6. Ice bucket to hold ZFM and to keep cells cool when placed in this medium.
7. Sterile plastic serologic pipettes (especially 1 ml and 10 ml) (Costar, Corning).
8. Pipetmen: P20 (1–20 μl volume), P200 (20–200 μl volume), P1000 (200–1000 μl volume)–(Rainin; tips—Robbins Scientific).
9. Sterile Pasteur pipettes (VWR)
10. Tubes: 12 × 75 mm sterile culture tubes with caps (Corning); 15-ml and 50-ml sterile conical centrifuge tubes (Corning).
11. Beakers to hold water and Alconox solution for cleaning chambers and to support 15-ml conicals at an angle in the incubator.
12. Racks to hold 12 × 75 tubes, 15-ml conicals, and 50-ml conicals. Styrofoam racks that are sold containing 15-ml sterile conical centrifuge tubes work well to support the helical chambers. The Styrofoam helps cushion the cells that are inside the chambers from rough movements. Have several available during fusion.
13. Sterile 96-well microtiter trays (Lindbro).
14. Trypan blue (.02%) for cell counts (Gibco).

III. MEDIA

A. Complete Growth Medium (10% FCS/IMDM)

Iscove's Modified Dulbecco's Medium (IMDM) (Gibco)

Fetal calf serum (FCS)

L-glutamine (100×29.2 mg/ml) (Gibco)

FM (100×): 0.45 g sodium pyruvate, 0.1 g bovine insulin, and 1.32 g cis-oxalacetic acid (Sigma). Dissolve in 100 ml double distilled water at room temperature with a stir bar (approximately 1 hr); sterile filter; freeze in aliquots at −20°C; freeze—thaw one time only.

2 Mercaptoethanol (2-ME)—(1000× stock) (Biorad). In fume hood, dilute 0.5 ml of 2-ME into 6.6 ml double distilled water; add 5 ml of this dilution to 95 ml double distilled water; sterile filter; freeze in aliquots at −20°C; thaw once, with subsequent storage at 4°C.

B. HT Medium

Complete growth medium with 15% FCS.

HT (100×): 0.776 g thymidine and 0.2772 g hypoxanthine (Sigma). Dissolve in 200 ml double distilled water at 70°C; sterile filter; store in aliquots at −20°C.

C. HAT/Ouabain 10⁻⁶ M

HT Medium

Aminopterin (1000×): 0.0176 g aminopterin (Sigma). Dissolve in 5 ml 1N NaOH; add 40 ml double distilled water; adjust pH to 7.0–7.3; bring volume to 50 ml; store in aliquots at −20°C.

Ouabain (10^6×): 0.0584 g ouabain (Sigma). Dissolve in 100 ml IMDM to yield 1000× stock; freeze in aliquots at −20°C; dilute 1:1000 just prior to use. Ouabain is light sensitive, so medium should be covered with aluminum foil.

D. Zimmermann Fusion Media

These media are available commercially from American Sterilizer Company (Erie, PA).

Mammalian cell fusion medium—#RD00-005

Mammalian cell postfusion medium-#RD00-006

Fusion Medium (ZFM)

Inositol (Serva #26310)–22.22 g (0.28 M)
K_2HPO_4 (Merck #5104)—74.85 mg (0.86 mM)
KH_2PO_4 (Merck #4872)—24.47 mg (0.36 mM)
Mg acetate (Merck #5819)—53.6 mg (0.5 mM)
Ca acetate (Merck #9325)—7.9 mg (0.1 mM)

Postfusion Medium (ZPFM)

NaCl (Merck #6400)—3.5 g (120 mM)
KCl (Merck #4936)—372.5 mg (10 mM)
K_2HPO_4 (Merck #5104)—740 mg (8.5 mM)
KH_2PO_4 (Merck #4872)—202 mg (3.0 mM)
Mg acetate (Merck #5819)—53.6 mg (0.5 mM)
Ca acetate (Merck #9325)—7.9 mg (0.1 mM)

For both media, dissolve dry ingredients in double distilled water; bring to 500 ml total volume; sterile filter the solution.

E. L3 Fusion Medium

Sorbitol (Merck #7759)—25.5 g (0.28M)
Mg acetate (Merck #5819)—53.6 mg (0.5 mM)

Ca acetate (Merck #9325)—7.9 mg (0.1 mM)

Pure BSA (Serva #11930)—0.5 g (1 mg/ml)

Dissolve dry ingredients in double distilled water; bring to 500 ml total volume; sterile filter. Use these sources of reagents to yield: pH 6.9–7.1 and osmolality 300 mOsm.

IV. CELLS FOR FUSION

All cells are cultured in a 37°C incubator with 6–$6\frac{1}{2}\%$ CO_2.

A. Mouse-Human Heteromyeloma Fusion Partners

Cells should be in log-phase growth with optimal viability. Dead cells do not migrate well in an electrical field and will interfere with the fusion process.

1. SBC-H20

Cells are maintained in horizontal flasks in 10% FCS/IMDM and expanded by splitting a relatively dense cell concentration of about 5×10^5/ml with 2–3 times the volume of fresh medium every other day. Change flasks once a week. Cells are fused 24–36 hr after feeding. Optimal fusion ratio is 1 B cell: 2–5 SBC-H20. Optimal DC voltage in ZFM is 60–70 DC volts (3–3.5 kV/cm) with most types of stimulated B cells. Cells will pellet well in growth medium when centrifuged at 110–$140 \times g$ for 10 min.

2. K_6H_6/B5

These cells are also maintained in horizontal flasks in 10% FCS/IMDM and expanded by splitting a relatively dense cell concentration (5×10^5/ml) with 2–4 times the volume of fresh medium every 3 days. Change flasks once a week. Cells are fused 36–48 hr after feeding. Optimal fusion ratio is 1 B cell: 3–5 K_6H_6-B5. This cell line is very heterogeneous. Optimal DC fusion voltage depends on the cells to which you are fusing, the cell number, and the fusion medium (30–90 V [1.5–4.5 kV/cm] range). Centrifuge cells in growth medium at 160–$200 \times g$ for 10 min to allow all the cells to pellet.

B. B Cells for Fusion

All fusions are performed with activated B cells.

1. In Vivo Activated B Cells

When B cells can be obtained from a patient during acute antigen challenge (e.g., acute infection), in vitro activation is generally not necessary. Peripheral blood lymphocytes are separated by Hypaque/Ficoll gradient centrifugation[15] and B cells isolated by rosetting with AET treated sheep red blood cells.[16] Antigen-specific hybrid yield is greater if the B cells are not frozen prior to fusion.

2. In Vitro Activated B Cells

The method of activation depends on the antigen system and the source of B cells. For example, two methods were used to produce the hybrids secreting antibody to HCMV:

EBV Activation. B cells are washed in 10% FCS/IMDM and resuspended in 30% FCS/IMDM containing 5–35% (v/v) supernatant from the marmoset line B95-8 as a source of EBV. If B cells are already activated somewhat in vivo, a lower concentration of marmoset supernatant is used. Resting B cells require a higher amount of EBV. Cells are plated in 96-well microtiter trays (Lindbro) at 10^5 cells/well and fed 30% FCS/IMDM twice a week. Cells are either fused before transformation is completed (at about 7 days), or after cells have become transformed and individual wells can be assayed for specific activity. Cells from reactive wells are then expanded or enriched (i.e., cloned at about 10^3 cells/well over fibroblast feeders, which have been irradiated with 10,000 rads) prior to fusion.

HCMV Antigen plus PWM Stimulation. Equal numbers of B cells and autologous helper T cells [separated by negative panning with Leu-2 antibody (Becton-Dickinson)[17] are cultured in 10% FCS/IMDM overnight in 96-, 48-, or 24-well trays (Costar) that have HCMV antigen coated on the wells. To coat the wells, HCMV antigen (Whitaker-MA BioProducts) diluted 1:1000–10,000 in PBS is incubated in the wells for 2 hr under ultraviolet light and rinsed out. Cells are added, and after an overnight incubation an equal volume of 0.1% PWM is added to each well. Activated B cells are fused 1–3 days later.

C. Feeder Layers

The need for feeder layers is dependent on the cell line and the fusion conditions. In general, the higher the voltage required to induce cell fusion, the more assistance the feeder layer will provide.

1. All SBC-H20 hybrids should be plated over feeders.
2. K_6H_6/B5 will give a high hybrid yield when fused to EBV-activated cells both with and without feeders. PWM-stimulated B cells used for fusion must be plated over feeders.
3. Feeder layer preparation:
   FS-3 (human foreskin fibroblasts) are grown to near confluence in the 60 inner wells of 96-well microtiter trays in 10% FCS/IMDM. IMDM containing penicillin/streptomycin (P/S, 100× stock at penicillin 10,000 u/ml and streptomycin 10,000 μg/ml [Gibco]) is placed in the outer wells to prevent evaporation. Prior to fusion (either the day before or earlier the same day), the plates are irradiated with 6,000 rads and replated with 1 drop HT in 60 inner wells, and IMDM containing P/S in the outer wells.

D. Hybrid Feeding Schedule

Day 1—Cells fused and plated in HT.

Day 2—Cells fed an equal volume of HAT/ouabain 10^{-6} M (2 drops/well with 5-cc serologic pipette).

Day 6—Cells fed HAT/ouabain 10^{-6} M (⅓–½ of media removed; fed 1 drop/well with 10-cc serologic pipette.

Day 10–Cells fed HAT/ouabain 10^{-6} M.

Day 14—If all cells in the control wells are dead, cells should now be fed with HT. At this point, some hybrids are visible macroscopically. Continue feeding twice weekly until supernatants are sufficiently spent from hybrid growth for assay.

METHOD I: *Fusions in the Helical Chambers with Zimmermann Fusion Media*

This procedure has been used successfully with SBC-H20 and B cells activated in vivo or in vitro with EBV or PWM (with or without antigen). Cells are fused with 1000 kHz, 5 V preset alignment for 30 s, followed by 60–70 V (DC). Although fusions have been successfully performed with a ratio as high as 1 B cell:10 SBC-H20, the more optimal ratio is 1 B cell:2–5 SBC-H20. With this fusion partner, a total of $5–11 \times 10^6$ cells has been fused successfully in a chamber; K_6H_6/B5 is also fused with this procedure. The DC voltage, however, is dependent upon the cells being fused. EBV-activated B cells will fuse at 60 V (DC), whereas PWM-stimulated cells require a higher voltage of 70–90 V (DC). A fusion ratio of 1 B cell:3–5 K_6H_6/B5 should be used with similar total cell numbers of $5–11 \times 10^6$.

1. Harvest the B cells and the heteromyeloma to be fused into separate conical centrifuge tubes. Throughout the procedure, conical tubes should be used to minimize cell loss during wash steps. Use the smallest tube that will hold the necessary volume. Activated B cells in culture are harvested either with a plastic serologic pipette or a Pipetman to avoid cell loss. Wells that contained the activated cells are rinsed with 10% FCS/IMDM to collect any residual cells.

2. Place 1–2 drops of each population of cells into separate 12×75 mm tubes with Pasteur pipettes to use as controls.

3. Centrifuge conical tubes at $160–200 \times g$ for 10 min. While cells are spinning, plate control wells:

   a. Add .3–.5 ml HT to each 12×75 mm tube containing cells.

   b. Plate at least 2 appropriately labeled wells with each population (1 drop with a 1 ml plastic serologic pipette).

   c. Plate 1 drop HT in at least 2 wells for feeder controls.

   d. Plate each population separately to ensure that the cells die in selection medium.

   e. Controls usually are not put through the wash steps.

4. Aspirate medium from the spun cells.

5. Resuspend to 10 cc 10% FCS/IMDM.

6. Add 50 μl cells to 450 μl trypan blue. Screw caps tightly on the tubes containing cells.

7. Count cells.

8. Calculate cell number of each population. Based on the number of available B cells, the type of activation, and the fusion voltage, decide on a plating density and a fusion ratio. Calculate the final volume of medium needed for plating.

9. Pool the cells to be fused in 15-ml conical tubes. If more than one fusion is to be performed with the same cells at the same fusion ratio, the cells can be pooled in one conical tube and washed together (not more than 15×10^6 cells total).

10. Centrifuge pooled cells at $250–300 \times g$ for 10 min. While cells are spinning, prepare on ice one 15-ml conical tube containing 10-ml cold ZFM for each population to be washed.

11. After centrifugation, aspirate the media to the cell pellets, and place the conicals on ice. Use a P1000 to resuspend each pellet with 1 ml of ZFM taken from the conical containing 10 ml. Mix the cells gently but well, making sure to end up with a single cell suspension. Overlayer the 1-ml cell suspension on the 9 ml of ZFM remaining in the 15-ml conical with the P1000 by pushing in the

plunger slowly and continuously while circling the tip around the inner walls of the conical just below the surface of the liquid interface.

12. Centrifuge cells at 10°C for 15 min at 140–300 × g. The harder spin will pellet the cells more effectively when their numbers are small. This is the fusion medium wash step. While the cells are spinning, aliquot 10 ml ZPFM into a 15-ml conical for each fusion to be performed. Place conicals containing medium in the 37°C water bath.

13. Aspirate medium off of washed cells to a "dry" pellet and place on ice.

14. Resuspend each pellet with 170–190 μl cold ZFM with a P200. The exact volume of medium to be added depends on the cell size and number. The chambers hold approximately 210 μl. If you have pooled cells for more than one fusion to wash together, resuspend to a volume that will allow you to divide them up among the appropriate number of chambers. Return each cell suspension to the ice with the conical cap on tightly.

15. One fusion at a time:

   a. Unwrap an autoclaved chamber and place parts in a Styrofoam rack, keeping the helix with the electrodes down and the receptacle with the open end up (Fig. 3-3).

   b. Transfer all of the cell suspension from the conical into the very bottom of the receptacle with a P200. Lift the receptacle out of the rack while you add the cells, so you can see that they end up on the bottom rather than along one side.

   c. Hold the receptacle in one hand and lift the electrode assembly out of the rack with the other. Invert the electrode assembly over the receptacle and insert it slowly inside the receptacle with a twisting motion, even though there are no screw threads. Once the helix has touched the cell suspension, be sure to keep moving it downward while you twist to prevent air bubbles from forming.

   d. Once the electrode assembly is firmly seated in the receptacle, wrap parafilm around their interface and return the chamber to the rack with the receptacle down and the electrodes sticking up.

   e. Take the rack containing the chamber to the power supply. It should already be turned on and preset as described in the equipment list.

      (1) Plug the connector cables leading from the power supply into the electrodes. Make sure they are firmly attached.

      (2) Push the alignment button on.

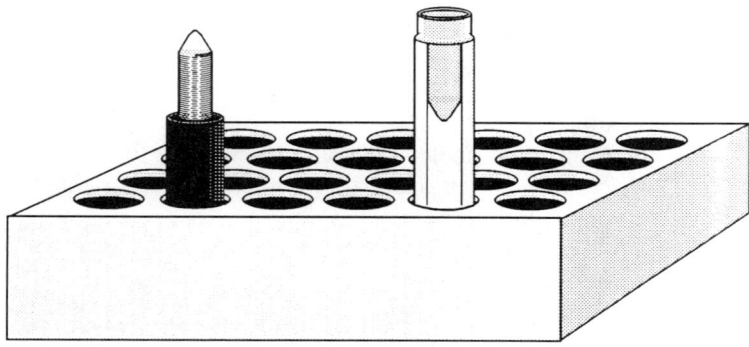

Fig. 3-3. Helical chamber in styrofoam rack.

(3) Quickly turn the alignment voltage adjust knob until the alignment voltage meter reads 5 V.

(4) After 30 s have elapsed on the process timer, push the DC cell fusion pulse trigger. With a 30-s preset alignment, the alignment voltage meter reading will gradually decrease to zero over the next 30 s.

(5) After the alignment has shut off, gently disconnect the electrodes and place the Styrofoam rack containing the chamber in the incubator for 10–20 min. It is very important not to disturb the cells with rough movements during either the fusion or postfusion incubation.

f. If more than one fusion is being performed, repeat all of Step 15 with the next fusion. Be sure to return the voltage alignment adjust knob back to zero before connecting the next chamber to the power supply.

16. After the chamber has been in the incubator for 10–20 min, harvest the cells into an empty 15-ml conical tube by rinsing the chamber with ZPFM prewarmed to 37°C.

a. Remove the helical assembly from the receptacle with a twisting motion. Replace the receptacle in the rack, but continue to hold the helical assembly with the electrodes up. There are cells in both the receptacle and in the helix that will need to be recovered.

b. With a P1000 set at 1000 μl, wash the cells out of the helix with ZPFM into the empty conical tube (Fig. 3-4). Hold the helical assembly at an angle over the conical so the liquid will drop down into the conical. The tip of the Pipetman should travel in parallel with the electrodes up and down the side of the helix facing you, and the helix should then be rotated in order to reach all surfaces. Use the same pipette tip and do two 1-ml washes in this manner—cover the helix thoroughly with the first 1-ml wash and then quickly with the second 1-ml wash, as a rinse.

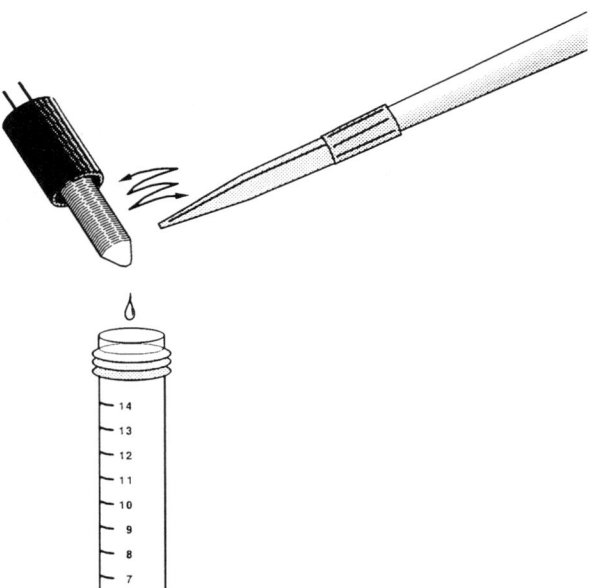

Fig. 3-4. Washing cells out of the helix.

    c. Use a plastic serologic pipette to wash cells out of the receptacle. Slowly add approximately 1½ ml of the remaining ZPFM from the prewarmed conical tube to the receptacle. Pipette gently up and down and transfer the cells to the same conical tube containing cells washed out of the helix. Avoid air bubbles. Repeat rinse of receptacle once or twice.

    d. Add remaining ZPFM to the conical tube containing the cells and pipette gently up and down.

17. Place 15-ml conical tube containing cells in the incubator at an angle (e.g., in a beaker) with the cap loose for 25–30 min.

18. The emptied helical assembly and receptacle are placed in a beaker containing Alconox and water for washing.

19. After the cells have incubated for 25–30 min, tighten the cap and centrifuge at $110–140 \times$ g for 10 min at room temperature.

20. Aspirate medium.

21. Using plastic serologic pipettes, resuspend cells in HT for plating in microtiter trays. Add 1 ml of HT with a 1-ml pipette while stirring slowly and pipetting gently up and down, and then add the remaining HT. Pipette gently up and down and plate the cells at 1 drop per well with a 5-ml pipette over a feeder layer previously plated with 1 drop of HT.

Points to Consider

1. Timing is critical. Adhere to incubation time ranges and do not leave cells in ZFM or ZPFM any longer than is necessary—2 hr maximum for ZFM and 1 hr maximum for ZPFM. As many as three fusions at one time have been performed successfully by this method.

2. The exact fusion ratio and plating density for each fusion are not determined until an actual cell count has been taken. Plating is usually done at $3–5 \times 10^4$ cells/well, but successful fusions have been performed plating at $2–9 \times 10^4$ cells/well. The higher plating density is generally used with more stringent fusion conditions (such as 90 V (DC)). Fusion ratios and plating densities are determined by a combination of experience with hybrid yield in earlier fusions, and observations made with new cell lines in the open fusion chamber (Method IV). For example, if the B-cell yield is 2×10^6 EBV-activated cells and fusions are to be done at 60 V (DC) with either K_6H_6/B5 or SBC-H20, 6×10^6 fusion partner cells would be a reasonable number to use (1:3 ratio). Because hybrid yield with EBV-activated fusions is generally high, these fusions are plated at $2–5 \times 10^4$ cells/well. Therefore, in a fusion of 8×10^6 total cells, 4 trays would be plated at 3.3×10^4 cells/well. If the B cells are activated with PWM and/or antigens, a higher voltage of 70 V (DC) or greater will be needed. The fused cells are then plated at a higher density of $5–9 \times 10^4$ cells/well. In the example given, 2 plates would be used at 6.6×10^4 cells/well.

3. To minimize cell loss, use the same pipette for a step that requires multiple manipulations (e.g., washing cells out of the receptacle).

4. Although this procedure was written with a 30-s preset alignment interval, the length of time can range from 20–30 s. The preset alignment time on the power supply should correspond with the number of seconds of alignment time prior to initiating the DC fusion pulse.

5. To calculate the final volume of HT needed for plating, consider 1 drop with a 5-ml plastic serologic pipette as approximately 60 μl. Approximately 3.6 ml is therefore needed to plate 60 inner wells of a microtiter tray. This volume

is for plating fusions in trays containing feeders already plated with 1 drop HT/well. If no feeders are used, either plate empty wells with 1 drop HT before beginning the fusion procedure or recalculate the volume needed for plating at 2 drops/well. Use only the 60 inner wells to plate cells. Place IMDM containing penicillin/streptomycin in outer wells to prevent evaporation from the inner wells.

6. As an indication of the effect of the DC voltage on the cells being fused and the consistency of the fusions, it has been useful to keep track of two fusion parameters:

   a. When the DC fusion pulse trigger is pushed, the alignment voltage meter reading usually goes up or down briefly. When there has been no movement at all, nothing has fused.

   b. A test of the alignment voltage can be made immediately after fusion (after the cells have been placed in the incubator). Before turning the alignment voltage adjust knob back to zero, push the alignment switch again. A number will appear on the alignment voltage meter. With SBC-H20 and K_6H_6/B5 fusions, it generally ranges from 10–22 V. The exact number depends on the number of cells in the chamber, the specific cells being fused, the DC voltage, and how well the cells are washed. Very high numbers outside this range indicate that everything is lysed or that nothing fused. A 5-V reading indicates that the chamber was not hooked up properly and no current got through.

METHOD II: *Fusions in the Helical Chambers with L3 Fusion Medium*

The fusion procedure with L3 fusion medium involves fewer washes and manipulations than the previous method. Consequently, fewer cells are lost in this procedure. All fusion steps are performed at room temperature. Fusions have been performed successfully with K_6H_6/B5 and EBV-activated B cells with a fusion ratio of 1 B cell : 3–5 K_6H_6/B5, 1–6 $\times$ 10^6 total cells per fusion, and the power supply set at 1000 kHz, 5-V preset alignment of 30 s, with 40–60 V (DC) for fusion.

1. Harvest the two populations to be fused in separate conical centrifuge tubes. Use the smallest tube that will hold the necessary volume. To avoid cell loss, harvest activated B cells in culture with either a plastic serologic pipette or a Pipetman. Rinse the wells that contained the cells with 10% FCS/IMDM to collect any residual cells.

2. Place 1–2 drops of each population into separate 12 $\times$ 75 mm tubes with Pasteur pipettes to use as controls.

3. Centrifuge conical tubes at 160–200$\times$ g for 10 min. While cells are spinning, plate control wells as in Method I, Step 3 a–e.

4. After centrifugation, aspirate the supernatant and resuspend the cells in 10 ml 10% FCS/IMDM. Take 50 μl cell suspension and add to 450 μl trypan blue. Close the caps tightly on the tubes containing the cells. Count the cells.

5. Place L3 fusion medium in 37°C water bath.

6. Based on the number of available B cells, the type of activation, and the fusion voltage, decide on a fusion ratio and plating density that will be appropriate. With K_6H_6/B5 and EBV-activated B cells, the plating density is 3–5 $\times$ 10^4 cells/well.

7. Pool both populations of cells in one 15-ml conical tube.
8. Centrifuge pooled cells at 250–300× g for 10 min.
9. Aspirate the medium.
10. Wash cells twice with 5 ml of L3 fusion medium. Take the L3 fusion medium from the water bath just prior to the first wash. After this step, the medium may be left at room temperature.

    a. Resuspend cell pellet with 1 ml L3 fusion medium using a P1000 Pipetman.

    b. Add 4 ml more L3 fusion medium with a plastic serologic pipette and gently mix by pipetting up and down.

    c. Centrifuge at 250–300× g for 10 min.

11. After the second wash, aspirate to a "dry" pellet.
12. Using a P200, resuspend the cell pellet with 170–190 μl L3 fusion medium. The exact volume to add depends on the cell size and number. The chambers hold approximately 210 μl.
13. Add cells to fusion chambers as in Method I, Step 15 a–d. If performing more than one fusion, all chambers should be filled before the actual fusion process begins. Place them in a Styrofoam rack with sufficient space between chambers. It is important that they not be moved once the fusion process has begun.
14. Take all of the chambers in the rack to the power supply. The power supply should be on and preset as described in the equipment list.

    a. Plug the connector cables leading from the power supply into the electrodes of the first chamber to be fused. Make sure that they are firmly attached.

    b. Push the alignment button on.

    c. Quickly turn the alignment voltage adjust knob until the alignment voltage meter reads 5 V.

    d. After 20–30 s have elapsed on the process timer, push the DC cell fusion pulse trigger. The preset alignment time should be the same as that used prior to initiating the DC pulse. The alignment voltage will gradually decrease to zero over that period of time.

    e. After the postfusion alignment has shut off, set the timer for 30 min. The incubation time after fusion is 30 min at room temperature.

    f. Gently disconnect the cables from the chamber. Do not move the chamber.

    g. If you wish to see the postfusion alignment, push the alignment button at this point. Record the voltage and push the alignment button again to shut it off. Turn the alignment voltage adjust knob back to zero.

    h. Repeat Steps a, b, c, d, f, and g with the next fusion.

15. Let the chambers sit motionless at room temperature for 30 min after the first chamber is fused.
16. During the 30-min incubation, prepare conical tubes containing HT with the correct amount of medium to plate each fusion. Fusions should be plated at 1 drop/well with a 5-ml plastic pipette (3.6 ml for the 60 inner wells of a tray).
17. After 30 min have elapsed from the end of the first fusion, wash the cells out of the helix and receptacle with the premeasured HT into an empty conical tube. Use the procedure described in Method I, Step 16 a–d.
18. Pipette gently up and down to mix.
19. Plate cells at 1 drop/well with a 5-ml plastic serologic pipette in trays containing feeders previously plated in 1 drop HT/well.

Points to Consider

1. Wash volume is reduced when fewer cells are fused (i.e., to fuse 2×10^6 cells total, 2–3 ml washes are used).
2. Because there is much less juggling of incubation times and temperatures, it is much easier to do several fusions at one time with this procedure. The limit to the number of fusions possible is more a function of the number of chambers available.
3. Cells should be left in L3 fusion medium no longer than 2 h maximum.

METHOD III: *Microfusions in L3 Fusion Medium (for 10^6 Cells or Less)*

Because the number of antigen-specific B cells is small, it is often desirable to fuse with very small cell numbers while the proportion of these cells is high. Fusions have been performed successfully using one microtiter well containing $1-3 \times 10^5$ EBV-activated B cells. Cells are fused at the same cell ratios as they are in larger fusions, but often on the high end with each heteromyeloma. For example, $K_6H_6/B5$ is used at the ratio of 1 B cell : 3–5 $K_6H_6/B5$, and if there are only 10^5 EBV-activated cells, 5×10^5 $K_6H_6/B5$ is used. Fused cells are plated at $3-5 \times 10^4$ cells/well. All equipment and reagents are the same as those used for larger cell numbers with L3 fusion medium except:

- The helical chamber is a Custom 200 μl Micro Helical Chamber with less dead space in the receptacle.
- Fusions are performed with a 20-s preset alignment interval. With very few cells, optimal electrical field strength is reached quickly.
- The DC voltage used is between 30–40 V. Because of the smaller cell number and the higher electrical field strength in a smaller receptacle, less voltage is required for fusion.

1. Because the number of B cells available is very small, and it is undesirable to lose a significant portion of them with a cell count, an estimation of cell number is made in one of two ways:
   a. A cell count from the well of cells to be fused using 10 μl of suspended cells in 1 ml trypan blue:
      (1) Estimate the volume of medium contained in the well by measuring the volume contained in a neighboring well of no interest.
      (2) Gently mix the cells to be fused with a P200 prior to removing 10 μl with a P20 for the cell count.
   b. Count cells contained in a different well from the same stimulation, which visually has about the same cell density, using 50 μl of cells in 450 μl trypan blue:
      (1) Measure the volume in the well of cells to be counted.
      (2) Mix cells to count gently with a P200 prior to taking 50-μl sample.
2. After counting the B cells, calculate their total number, the number of cells of the fusion partner needed, and the final volume of HT needed to plate the cells at the desired cell density. Cells are plated in microtiter trays at 1 drop/well with a 1-ml plastic serologic pipette over feeders previously plated in 1 drop HT. One drop with a 1-ml plastic serologic pipette is approximately 50 μl.

3. Harvest heteromyeloma into a conical centrifuge tube. Add 1–2 drops with a Pasteur pipette to a 12 × 75 mm tube to use as a control.

4. Centrifuge the heteromyeloma at 160–200× g for 10 min. While it is spinning, plate the controls (at least 2 wells of each cell population):

   a. Fusion partner control—identical to Methods I and II.

   b. EBV-activated cell control—cells from a different well of the same stimulation are used, preferably a well containing cells similar in size, state of activation, and secreting a small amount of antibody. Add 1–2 drops of cell suspension from the control well and .3–.5 cc HT to a 12 × 75 mm tube. Plate at 1 drop/well.

5. After the centrifuge has stopped, aspirate the supernatant from the fusion partner. Resuspend in 10 ml 10% FCS/IMDM and count cells using 50 μl of cells and 450 μl trypan blue.

6. Add the number of fusion partner cells calculated in Step 2 to a 15-ml conical.

7. Use a P200 to harvest the well of activated B cells for fusion, and add the cells to the conical containing the fusion partner. Rinse the well several times with 10% FCS/IMDM.

8. Centrifuge at 250–300× g for 10 min and aspirate the supernatant.

9. Wash cells twice with 1 ml of L3 fusion medium. The fusion medium should be taken from the 37°C water bath just prior to the first wash, and left at room temperature for further use. Add the L3 fusion medium for the washes with a P1000. Spin at 250–300× g for 10 min.

10. After the second wash, aspirate to a "dry" pellet and resuspend cell pellet in 150–160 μl L3 fusion medium. Add entire volume to chamber receptacle, screw in helix, and parafilm the interface. (See Method I, Steps 15 a–d.)

11. Fuse cells by the procedure used for larger cell numbers in L3 fusion medium, using 20s preset alignment, and 30–40 V (DC) for fusion (Method II, Steps 14 a–h).

12. After 30 min incubation at room temperature, wash cells from the helix and receptacle with the amount of premeasured HT needed to plate the fusion. Because the amount of medium available to wash cells out of the helix and receptacle is much smaller, the steps used are similar to those described in Method I, Steps 16 a–d, with the following changes:

   a. Using a P1000 with 500 μl HT, wash the helix out into the receptacle, rather than an empty conical tube. Wash once more with 500 μl HT.

   b. Gently pipette up and down in the receptacle with a 1-ml plastic serologic pipette and transfer the cells to an empty conical tube.

   c. Rinse the receptacle with the remaining HT (using the same pipette) and add to the conical tube containing the cells.

   d. Plate cells at 1 drop/well with the same 1-ml pipette over feeders already plated in 1 drop HT.

METHOD IV: Fusions in the Open Chamber

The open chamber allows one to establish fusion parameters for a particular cell population by actually observing the fusion process. Conditions are not identical to those in the helical chambers, but they are sufficiently close to establish an appropriate range to start. Ideally, the goal is to determine suitable electrical field strength for each cell population, fusion partner, and activated B cells, separately. Based on this information, use a range between the two to begin the optimization of fusion parameters. In general (but not always):

> Cells will align with 1000-kHz frequency and 5-V AC current if they are washed free of serum protein.

> DC voltage needs to be higher with smaller and less activated cells, and lower with larger and more activated cells.

Work with one population at a time to begin the process. Following are the steps to take.

1. Count the cells before you start. A cell density of $1-3 \times 10^7$ cells/ml will be roughly equivalent to a fusion of $2-6 \times 10^6$ cells in the helical chamber. A higher cell density, especially with large cells, will make it very difficult to observe cell fusion in the chamber. A lower cell density will change the translation of the voltage parameters to higher cell numbers.
2. Centrifuge the cells and wash with fusion medium as you would in preparation for fusion in the helical chambers. Washes must be sufficient to remove serum proteins. Care must be taken not to lose cells by centrifuging too gently, using too large a volume of wash medium, or centrifuging too hard.
3. After washes, resuspend cells in fusion medium at $1-3 \times 10^7$ cells/ml (on ice for ZFM and room temperature for L3 fusion medium).
4. Add 14 μl of the cell suspension to the open chamber by pipetting it to the center of the two electrodes with a P20. Coverslip. Avoid air bubbles.
5. Place the chamber under a microscope (either a regular or an inverted microscope will work, but it must have a slide holder to keep the chamber from moving). Plug the electrodes into the connector cable from the power supply. Tape the connector cable to the microscope if there is any chance of moving it. Focus the microscope on the electrodes under high magnification. If the cells are wicking out between the electrodes, add small drops of cell suspension to the edges of the coverslip at the electrodes.
6. The first parameter to observe is the alignment. Before the cells can be fused by a DC pulse, they need to be lined up in chains between the electrodes with flattened areas at their cell membranes where the cells are in close contact with one another (Fig. 3-5). This flattened area is where the fusion will occur. Using higher AC current than necessary will lead to fewer cells remaining viable through the fusion process. Not enough AC current will cause the cells to rotate and fail to form the requisite membrane contact between cells.

   Start with 1000 kHz frequency. Push on the alignment switch and slowly turn the AC voltage adjustment up to 5 V. Observe at what voltage the cells start to line up. Are they lining up smoothly, or are there uneven movements?

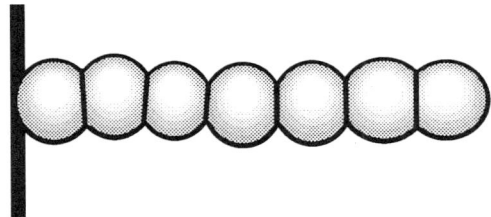

Fig. 3-5. Cells aligned with AC current.

Cells not washed sufficiently or with very poor viability will not align well. If the cells seem to be aligning at 1000 kHz, 5 V (AC), take a fresh 14-μl cell suspension and put it in a new chamber. After hooking up the chamber, push the alignment switch on and quickly turn the AC voltage adjustment up to 5 V, as you would with the helical chamber. If the cells are lined up with flattened adjacent cell membranes within 20–30 s, the alignment is satisfactory. Cells will line up faster with 2000 kHz frequency and/or 6–10 V. However, both heteromyelomas and activated human B cells will align at the lower settings provided they are thoroughly washed and their viability is high.

7. The second parameter to determine is the DC voltage needed for fusion. With the mouse-human heteromyelomas, it is in the 40–70 V (2–3.5 kV/cm) range. Set the power supply at 30-s postfusion alignment and set a DC voltage with the fusion adjust knob. Place a fresh 14-μl cell suspension into a new chamber. Hook up the chamber. Push the alignment switch on. Quickly turn up the AC voltage adjust to 5 V. After 30 s have elapsed on the process timer, push the DC voltage cell fusion pulse trigger.

Watch what is happening to the cells under the microscope. Fusion does not take place immediately. You will be observing intermittently for 20–30 min. It is important that the chamber not be removed from the microscope stage during this period. Change fields slowly and smoothly so as not to disturb the cells. Generally, when the DC fusion pulse is introduced, there will be a slight jostling of the cells. The higher the DC voltage used, the bigger the jostle. If the voltage is too high, many cells will lyse or start to lose membrane integrity immediately. What you are watching for is the fusion of cells (Fig. 3-6).

Ideally, in a very homogeneous population (which mouse-human heteromyelomas are not), you will see chains of fused cells formed (Fig. 3-7) within 20–30 min. Cells will often appear very grainy and then recover. If you see no cells fusing, try higher DC voltage. If they are disintegrating, try lower voltage. Observe the health of the cells each time you load fresh cells into a new chamber. Some cells will remain healthy in fusion medium longer than others. If a particular cell line is not maintaining its viability well, do not try to do more than one fusion at a time in the helical chambers. This will minimize the time the cells are exposed to fusion medium. The DC voltage required for fusion depends on the cell, its size, its shape, its state of activation, its health, the stage of its growth cycle, the fusion medium, and so on.

Once the optimum fusion parameters for each cell populations are determined, two cell populations can be fused at once. Following is an outline of the process.

1. Wash each cell population separately. Pool cells in different ratios to work with in the open chamber. Different DC voltages work better with different cell ratios. Start at a 1:1 ratio to get a range. Observe which population is lost when the

Fig. 3-6. Two cells fusing.

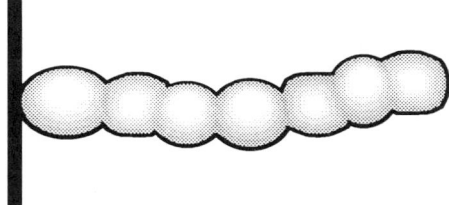

Fig. 3-7. Chain of fused cells.

voltage becomes too high. In general, a relatively high DC voltage is used to insure fusion of the activated B cells. This intensity, however, will tend to lyse significant numbers of heteromyelomas. The loss is tolerated because fusion of activated B cells is the goal and their number is limited. For example, one B-cell population activated with 0.1% PWM for 2 days fused best to itself at 100–110 V (DC) in ZFM; K_6H_6/B5, alone, fuses best between 40–70 V (DC). The voltage used to fuse these B cells with K_6H_6/B5 was 70–90 V (DC).

2. The goal of the process is to establish the optimum parameters for the maximum number of hybridomas of one B cell fused to one heteromyeloma. Because the cell populations differ, they will tend to migrate to the electrodes at different rates rather than randomly distributed. This often leads to two cells of one population fusing with a third cell of the other population. With uneven cell ratios this tendency is increased, leading to the formation of unstable hybrids.

Points to Consider

1. It is important to have several open chambers available for use at the same time. A chamber should not be reused until it has been washed and dried.

2. The length of time it takes cells to align depends on how well the cells are washed, their number, viability, size, and such. For fusions with small numbers of cells (about 10^6 cells), 20 s of alignment are adequate. For larger numbers of cells, the alignment time is more dependent on the specific cells being fused. Observations of alignment time in the open chamber are helpful but should be extended to the helical chambers. If many giant cells are observed postfusion, it is likely that more than 2 cells are fused. Decreasing the alignment time will reduce their number.

3. The open chamber allows visualization of the fusion process, but there are some limitations that must be taken into account:

   a. The time in fusion medium is critical. The voltage needed to cause fusion will change if the cells have been in the medium for a long time, so it is important to adhere to the 2-hr time limit. In practical terms, this means there is usually time for only 3 or 4 rounds in the open chamber with one cell population.

   b. Mouse-human heteromyelomas are hybrids themselves. In low ionic fusion media, they will often appear like two cells fusing, when they are actually only one cell. Confirm that this phenomenon is two different cells fusing, not just the fusion partner reacting to the medium.

   c. With some stimulated B cells, the B cell cannot always be distinguished from the fusing partner, so they have to be recognized by their behavior. For example, K_6H_6/B5 is so heterogeneous that no more than an occasional pair or triplet will be formed fusing to itself with different DC fusion voltages and cell densities. When K_6H_6/B5 is mixed with B cells, however, many more pairs, triplets, and some longer chains will be formed. This becomes a very useful observation when fusing with EBV-activated B cells that often cannot be visually distinguished from K_6H_6/B5.

4. Because cell lines change constantly, no two lines are exactly alike, and the number of B cells secreting the antibody of interest is limited, it is important to gain experience with the fusion partner and a variety of stimulated B cells in the open chamber to optimize fusions in the helical chambers.

V. CONCLUSIONS

The ability to produce human hybridomas with a relatively small number of input B cells will facilitate wider applications of human monoclonal antibodies in research and ultimately in the treatment of human diseases. Electric field-induced cell fusion or electrofusion provides a tool to achieve this goal with the advantages of increased fusion efficiency and the ability to perform microfusions. The technique allows the investigator to manipulate the fusion parameters specific to the fusion partner and the B cells being immortalized which, in general, has not been possible with chemical fusogens. For a variety of reasons, the production of human hybridomas has been technically difficult. There is a great heterogeneity among human B cells and B-cell lines. Optimal growth conditions and cell fusion parameters developed with one population may not necessarily succeed with another. Electrofusion parameters, however, can be changed to achieve cell fusion with any cell population. Because all cells being fused are not alike, fusion parameters such as cell ratio (ratio of B cells to fusion partner), AC voltage alignment time, and DC voltage can be manipulated to optimize the fusion of one B cell to one fusion partner. To achieve consistent fusion efficiency with a particular fusion partner, it is important that the care and feeding of this cell line be consistent. This is particularly true for mouse-human heteromyelomas, which are hybridomas themselves and will vary according to their growth conditions and longevity in culture.

A sufficient source of antigen-specific B cells is prerequisite to the formation of hybridomas secreting specific HMAbs. In general, these cells are scarce, difficult to maintain in culture, and quite heterogeneous. In our experience, there are sufficient differences among individuals that we rarely expect to produce relevant hybridomas from one fusion event. A systematic and technically consistent approach to develop B-cell activation schemes and fusion parameters is needed. To accomplish this task in this phase of the work, it is best to procure and cryopreserve as many cells as possible from the human donor at a specific time. When in vivo activated B cells are not available, in vitro activation is needed to stimulate the B cells to reach the appropriate stage of development to form Ig-secreting hybridomas and to expand the desired antigen-specific B-cell population. The specific activation is dependent upon the antigen system and the source of these B cells. Because B cells must be fused at the optimal time in their growth and activation, a rapid and consistent assay for specific antibodies must be available. Hybrids must be cloned to isolate stable Ig secreting hybridomas and to ensure monoclonality.

In summary, electrofusion is a technique for manipulating the cell fusion process itself. In order to be successful in the generation of human hybridomas secreting relevant antibodies, however, it is important that the investigator obtain or develop a source of activated B cells primed to secrete relevant antibodies, become familiar with the growth patterns of the cells he or she is fusing in order to maximize the number of hybridomas that can be produced, and screen for antibody-secreting hybridomas quickly in order to preserve them.

ACKNOWLEDGMENTS

The authors wish to thank Judy Halmos for technical assistance and Judy Campbell, who typed this manuscript. This study was supported in part by grants HL33811, AI22557, and AI26031 from the National Institutes of Health and grants from the Deutsche Forschungsgemeinschaft (SFB 176, 135) and the BMFT (01QV354).

REFERENCES

1. James, K., and Bell, G. T. 1987. Human monoclonal antibody production, current status and future prospects. J. Immunol. Meth. 100:5–40.
2. Foung, S. K. H., Perkins, S., Raubitschek, A., Larrick, J., Lizak, G., Fishwild, D., Engleman, E. G., and Grumet, F. C. 1984. Rescue of human monoclonal antibody production from an EBV-transformed B cell line by fusion to a human-mouse hybridoma. J. Immunol. Meth. 70:83–90.
3. Foung, S. K. H., Blunt, J. A., Wu, P. S., Ahearn, P., Winn, L. C., Engleman, E. G., and Grumet, F. C. 1987. Human monoclonal antibodies to Rh_oD. Vox. Sang. 53:44–47.
4. Foung, S. K. H., Blunt, J., Perkins, S., Winn, L., and Grumet, F. C. 1986. A human monoclonal antibody to rh^G. Vox Sang. 50:160–163.
5. Alpert, S. D., Turek, P. J., Foung, S. K. H., and Engleman, E. G. 1987. Human monoclonal anti-T cell antibody from a patient with juvenile rheumatoid arthritis. J. Immunol. 138:104–108.
6. Foung, S. K. H., Perkins, S., Koropchak, C., Fishwild, D. M., Witteck, A. C., Engleman, E. G., Grumet, F. C., and Arvin, A. M. 1985. Human monoclonal antibodies neutralizing varicella-zoster virus. J. Infect. Dis. 152:280–285.
7. Foung, S. K. H., Perkins, S., Arvin, A., Lifson, J., Mohagheghpour, N., Fishwild, D., Grumet, F. C., and Engleman, E. G. Production of human monoclonal antibodies using a human-mouse fusion partner. In Human Hybridomas and Monoclonal Antibodies, E. G. Engleman, S. K. H. Foung, J. Larrick and A. Raubitschek, eds., Plenum Press, New York, 1985.
8. Banapour, R., Rosenthal, K., Rabin, L., Sharma, V., Young, L., Fernandez, J., Engleman, E., McGrath, M., Reyes, G., and Lifson, J. 1987. Characterization and epitope mapping of a human monoclonal antibody reactive with the envelope glycoprotein of human immunodeficiency virus. J. Immunol. 139:4027–4033.
9. Carroll, W. L., Thielemans, K., Dilley, J., and Levy, R. 1986. Mouse × human heterohybridomas as fusion partners with human B cell tumors. J. Immunol. Meth. 89:61–72.
10. Foung, S. K. H., Perkins, S., Bradshaw, P., Rowe, J., Rabin, L. B., Reyes, G. R., and Lennette, E. T. 1989. Human monoclonal antibodies to human cytomegalovirus. J. Infect. Dis. 159:436–443.
11. Foung, S. K. H., and Perkins, S. 1989. Electric field-induced cell fusion and human monoclonal antibodies. J. Immunol. Meth. 116:117–122.
12. Zimmermann, U. 1986. Electrical breakdown, electropermeabilization and electrofusion. Rev. Physiol. Biochem. Pharmacol. 105:175–256.
13. Zimmermann, U. Electrofusion of cells. In Methods of Hybridoma Formation, A. H. Bartal and Y. Hirshaut, eds., Humana Press, New Jersey, 1987.
14. GCA/Precision Scientific Group. Instruction and Protocol Manual, Zimmermann Cell Fusion System, GCA Corporation, Chicago, Illinois, 1983.
15. Foung, S. K. H., Perkins, S., and Engleman, E. G. Peripheral blood lymphocyte separation from whole blood or buffy coats. In Human Hybridomas and Monoclonal Antibodies, E. G.

Engleman, S. K. H. Foung, J. Larrick, and A. Raubitschek, eds., Plenum Press, New York, 1985.

16. Foung, S. K. H., Coutre, S., and Engleman, E. G. Separation of human T and non-T lymphocytes from peripheral blood. In Human Hybridomas and Monoclonal Antibodies, E. G. Engleman, S. K. H. Foung, J. Larrick, and A. Raubitschek, eds., Plenum Press, New York, 1985.

17. Coutre, S., Benike, C. J., and Engleman, E. G. Panning for human T-lymphocyte subpopulations. In Human Hybridomas and Monoclonal Antibodies, E. G. Engleman, S. K. H. Foung, J. Larrick, and A. Raubitschek, eds., Plenum Press, New York, 1985.

CHAPTER 4

Generating Immortalized Immunoglobulin-secreting Human Lymphocytes by Recombinant DNA Technology

Martin I. Mally and Mark C. Glassy

INTRODUCTION

The classical technique of producing hybridomas involves the fusion of B lymphocytes, albeit at a low efficiency (approximately 1 hybrid per 10^5–10^6 cells), with an appropriate fusion partner, typically a myeloma or lymphoblastoid cell line, via the cumbersome and somewhat unpredictable polyethylene glycol (PEG) procedure.[1] The fusion partner simply acts as a vector, supplying immortalization functions to an immunoglobulin-secreting lymphocyte. Recently, however, other approaches have become available that have the potential to generate immortalized, immunoglobulin-secreting lymphocytes with a higher efficiency. The aim of this chapter is to provide an introduction to an alternative method of generating immortalized lymphocytes—that of a form of DNA-mediated gene transfer known as electroporation.

Recombinant DNA technology has afforded a novel approach to the generation of immortalized, immunoglobulin-secreting lymphocytes. Several methods exist in which exogenous DNA is introduced into cells permanently. These include gene transfer, or transfection, by calcium phosphate-DNA coprecipitation[2,3,4,5]; DEAE dextran neutralization[6,7]; the use of viral vectors,[8,9] liposomes,[10,11,12] or protoplast fusion[5,13-17]; and direct microinjection of DNA into the nucleus of recipient cells.[18-20] Gene transfer can also be accomplished by the brief exposure of cells to a high-voltage electric pulse.[21-23] This technique, termed electroporation, uses a transient electrical direct current (DC) to temporarily and reversibly permeabilize the cellular membrane, facilitating the uptake of exogenous materials such as DNA.[22,24,25] Electroporation is rapidly becoming the method of choice for transfecting lymphocytes, as they are quite refractory to most of the other gene transfer techniques.[15,24,26-28] When cultured cells are electroporated in the presence of DNA, the DNA enters the cells and is randomly and, in certain cases, stably integrated in the genome of a select number of cells. If selective pressure is applied, for example by culturing the cells in the presence of a cytotoxic drug or antibiotic, the subset of cells that has incorporated and is expressing the gene conferring resistance to that particular chemical will continue to survive, similar to the method by which hybridomas are selected in HAT media. Conversely, cells that have not incorporated the drug resistance gene will not survive. This enables the investigator to obtain a very pure population of cells with the desired characteristics. Thus, one can envision the generation of hybridoma-like cells from activated B cells by electroporation in the presence of DNA sequences coding for immortalization functions. These sequences are readily available to the scientific community and will be discussed later. The term *electrodoma* is introduced here, and refers to the immunoglobulin-secreting lymphocytes resulting from an electroporation-mediated insertion of immortalizing DNA.

A number of electrical and biological parameters must initially be optimized in order to achieve high-efficiency electroporation. These include the electric field strength, pulse width, electroporation medium, cell type, DNA conformation and concentration, and temperature. Each of these variables will now be discussed in more detail.

Electroporation involves applying a brief, well-controlled DC electric field between two electrodes surrounding cells in suspension. Transient pores are created within the membrane, facilitating the entry of macromolecules into the cell.[21,22,29,30] Neither the structural characteristics of the pores nor the biophysical mechanisms of the induction of DNA uptake are well understood. It appears that the voltage applied to the cells in the electroporation chamber is the most critical parameter affecting high-efficiency gene transfer. Since mammalian cells differ in size and membrane composition, the optimal electric field strength must be experimentally determined for each different cell type being examined.[22-24,26-28,31-35]

The electric field strength, expressed in volts/centimeter (V/cm), is defined as the voltage delivered divided by the distance between the electrodes. Two different types of electric pulses have been used to electroporate cells: square wave and exponential decay. The voltage for a square wave is adjusted to a given amplitude, maintained for a specified time period or "pulse width," and is then returned to zero. The voltage of an exponential pulse (the less-expensive method) is raised to an initial peak amplitude and then decays as an exponential function of the resistance and capacitance. In fact, square wave pulses have not been found to be more effective in electroporating cells than the capacitor discharge method. Therefore, only the exponential waveform will be discussed. The exponential decay waveform is generated by the discharge of a capacitor and can be characterized by three variables: the size of the capacitor (i.e., the amount of charge the capacitor will store at a given voltage, usually expressed in microfarads), the voltage to which the capacitor is charged, and the resistance of the medium through which the capacitor is discharged. The length of time required for decay of the electric field generated during electroporation is described by a decay constant, τ (also termed the resistance-capacitance or RC time constant), as the time necessary to reduce the initial voltage to $1/e$ ($\sim 37\%$) of the maximal discharge voltage; τ therefore depends on the size of the capacitor and the resistance through which it is discharged. A larger capacitor or a higher resistance medium requires a longer time for the capacitor to discharge.

The time constant in an electroporation experiment can be adjusted by altering the size of the capacitor and/or the resistance across the electroporation chamber. The size of the capacitor is limited by the electronic specifications of the electroporation apparatus. The resistance within the sample chamber is affected by the ionic strength of the electroporation buffer and the cuvette geometry. As ionic strength increases, the resistance of the buffer decreases. Therefore, a pulse delivered into a buffer of higher ionic strength will have a shorter time constant, assuming all other variables (i.e., voltage, capacitance, cuvette geometry, volume) remain constant. Alternatively, changing the resistance in the electroporation chamber can be accomplished by varying the volume of buffer in the chamber. The resistance is proportional to the length of the path through the buffer and to the cross-sectional area of the path. The path length (or interelectrode distance) in the chambers is fixed, usually at 2–5 mm. The cross-sectional area, however, changes with the volume of the sample, which in turn alters the resistance and the time constant. Increasing the interelectrode distance or decreasing the volume in the chamber effectively increases the resistance, thereby increasing the time constant.

The electroporation chamber contains cells, growth medium, and the DNA to be transfected in suspension between two uniformly spaced electrodes. A variety of chambers is now commercially available, with different geometries, volumes, and electrode spacings. The exact geometry of the chamber is probably not as important as the generation of a suitable electric field within the chamber. The configuration of the chamber should allow a uniform electric field to be delivered throughout the chamber to allow all of the cells to be exposed to the same electric environment. Cell toxicity due to electrode corrosion can be a problem for aluminum or stainless steel electrodes. Therefore, it is convenient to use disposable or easily sterilizable chambers, typically with a maximum capacity of 1–1.5 ml. Many of the current electroporation instruments use commercially available spectrophotometer cuvettes fitted with electrodes as sample chambers.[24,26,27]

The choice of an appropriate electroporation medium must also be considered, because it influences cell viability and transfection efficiency in several ways. The electric field generated across two fixed flat-plate electrodes is described by Ohm's law, $V = IR$, where V represents the voltage, I is the current, and R is the resistance. Thus, the resistance, or conductivity, of the medium in which the cells are suspended has major effects on the current generated. Altering the ionic strength of the electroporation medium will alter the electric field and change the optimal voltage for electroporation. Since electroporation requires a combination of electric field strength and time constant, it is important to consider the effects of ionic concentration and osmolarity on the time constant. The osmotic balance of the electroporation medium can be maintained with salt, or with nonconductive components such as sucrose or other nonionic molecules. The ionic strength of the medium affects sample resistance as well as the formation of the transmembrane pores. Higher salt concentrations correspond to lower resistance, which leads to a shorter pulse time and therefore to smaller pore sizes.[36,37] This in turn affects DNA transfection and cell viability.

Buffered saline solutions and tissue culture medium have been used successfully in many electroporation studies.[27,34] Both are low-resistance solutions (i.e., high conductivity) compared to iso-osmolar sucrose solutions. In a low ionic strength solution, such as 0.1 M sucrose, modest voltages will generate an electric field for electroporation. However, many mammalian cell types, especially hematopoietic cells, may be extremely sensitive to nonphysiological solutions, thereby adversely affecting their viability.[27] In any case, the electroporation buffer chosen should be one in which the cells thrive and retain high viabilities (i.e., tissue culture medium with or without added serum, buffered saline solutions).

The condition of the cells is a critical factor in obtaining high-efficiency electroporation. Cells should be grown under optimal culture conditions and transferred to fresh media just prior to electroporation. The cells should preferably be in log-phase growth at the time of the experiment, since recovery of stable transfectants is better when actively growing cells (in comparison to quiescent cells) are electroporated.[24,38] Replicating cells appear to have an increased ability to take up and integrate exogenous DNA than resting cells. If fully confluent or G_0/stationary phase cells are used for electroporation, a two- to tenfold reduction in efficiency was observed.[26] When cells were incubated in the presence of colcemid for 24 hr prior to electroporation (which arrests cells at the mitotic phase), an increased transfection efficiency was obtained, suggesting that mitosis may be a period of enhanced competence for DNA uptake.[24] Different cell types have poorly understood variations in their innate abilities to take up DNA by electroporation. For example, cells grown in suspension usually require significantly greater field strengths for optimal electroporation than adherent cells.

DNA conformation (supercoiled versus linearized) and concentration also influence transfection efficiency. Many reports have appeared in the literature demonstrating that linearized DNA was two- to twentyfold more efficient at yielding stable transfectants via electroporation than supercoiled DNA.[24,26,28] This may reflect either a higher recombinational activity of the linearized DNA with the host genome, or a decreased efficiency for supercoiled DNA to reach the nucleus.[28] The DNA integrates at random sites throughout the genome, apparently with few if any mutations.[26,39] Increasing the concentration of linearized DNA (typically, 1–20 μg/ml) resulted in increased numbers of stable transfectants.[22,26,34,38,40] It should be mentioned that large numbers of cells can be transfected with relatively small amounts of DNA (in the microgram range). Electroporation of cells resuspended at concentrations ranging from 2.5×10^6 to 4×10^7 per ml showed no significant differences in electroporation efficiency.[26]

The ability of a given cell type to express a particular DNA sequence is also important in determining the amount of DNA used for electroporation.[27] The size of DNA successfully transfected ranges from small plasmids (i.e., 3–6 kb) to high molecular weight DNA (150 kb), with little change in the transfection efficiency.[27,41] The simultaneous transfection of two separate plasmids containing different selectable markers into electroporated cells has also been achieved. However, the efficiency of cotransfection is higher than would be expected from the random uptake of two plasmids. For example, it has been reported that as many as 100% of cells transfected with one plasmid also took up a second different plasmid.[28,40,42] The results of these experiments suggest that a subpopulation of competent cells exists which is capable of integrating transfected DNA into their genome.

When a capacitor discharges its electric energy into a resistant medium, it generates heat. Therefore, it seems likely that increased cell viability and stabilization of pore formation would occur at 4°C rather than at 23°C.[24,38] It is well known that temperature affects the physical properties of membranes, and it certainly influences the duration of the permeabilized state. However, it has been found empirically that performing electroporation and all subsequent manipulations at room temperature is more effective for stable transfection.[26] Although electroporation at 4°C does enhance cell viability, absolute efficiency appears to be two- to fivefold better at 23°C than at 4°C. Electroporated cells may remain permeable for several hours if low temperatures are maintained. Loss of permeability is retarded when cells are maintained at 4°C, but occurs within a matter of minutes when cells electroporated at 4°C are subsequently incubated at 37°C.[43] Therefore, as a general rule, electroporation of mammalian cells should be performed at room temperature.

To achieve efficient stable transformation of mammalian cells by DNA transfection, experimenters must consider two factors: efficient delivery of DNA into the cell nuclei to promote its integration into the host chromosome (as discussed earlier), and sufficient amounts of expression of the transduced gene to allow the cell to survive and grow in the selection medium. Random integration of prokaryotic genes into eukaryotic cellular DNA and their subsequent expression under the control of eukaryotic promoter regions is a rare event (10^{-7} to 10^{-8}). An initial decision must therefore be made concerning the method of selecting the recombinant electrodomas, since only a minority of cells is stably transfected. When cells are electroporated in the presence of a plasmid containing a gene encoding for a drug resistance marker, those cells that incorporate the DNA into their genome will become resistant to that drug.

Probably the most commonly used dominant acting selectable gene (i.e., a gene that has no mammalian counterpart, thereby precluding the need for genetically mutant cells) is the bacterial neomycin resistance gene. This gene is of bacterial origin and is normally nonfunctional in mammalian cells. However, ligation of an eukaryotic promoter at the 5′ end permits its expression in mammalian cells.[44,45] The bacterial transposon Tn5 encodes a gene (neo) whose protein product is an aminoglycoside 3′-phosphotransferase type II, which confers resistance to the kanamycin-neomycin group of antibiotics. Mammalian cells are normally resistant to neomycin, but are sensitive to one of its derivatives, so-called G418 or geneticin.[46,47] This aminoglycoside antibiotic is toxic to most eukaryotic cells by virtue of its interference with the function of the 80S ribosomes and the subsequent inhibition of protein synthesis. Mammalian cells that have acquired and expressed a functional neo gene by DNA-mediated gene transfer would therefore become resistant to G418 toxicity and survive in the presence of the drug.

Another widely used bacterial gene that also has no counterpart in mammalian cells is the gene (gpt) encoding for the enzyme xanthine-guanine phosphoribosyltransferase (XGPRT or XPRT). This enzyme is active in the de novo synthesis of the purine GMP (guanosine monophosphate). Once again, coupling of an appropriate promoter at the 5′ end of the prokaryotic gene allows its expression in eukaryotic cells.[48,49] The natural substrate of XGPRT is xanthine, although it can also use hypoxanthine, but less effectively. By contrast, the mammalian hypoxanthine-guanine phosphoribosyltransferase does not utilize xanthine efficiently as a substrate. In fact, mammalian cells do not convert xanthine to xanthylic acid (XMP) or guanylic acid (GMP) at a significant rate. The success of this selection protocol depends upon the use of metabolic inhibitors. Mycophenolic acid specifically inhibits IMP dehydrogenase, the enzyme responsible for the conversion of IMP to XMP, which subsequently is converted to GMP.[50] Aminopterin blocks the de novo synthesis of IMP from its precursors. Supplementing the medium with hypoxanthine or adenine cannot overcome the inhibition caused by these two compounds. The addition of guanine and either hypoxanthine or adenine can reverse the inhibition of cell growth, however, because these bases can be converted to their respective mononucleotides by the purine salvage pathway. Growth of transfected cells in a medium that contains xanthine but not guanine enables cells to synthesize GMP from xanthine via expression of the bacterial XGPRT. Therefore, the use of agents that inhibit purine nucleotide synthesis (i.e., mycophenolic acid and aminopterin) together with the provision of xanthine as the sole source of purines make it an excellent selection system.

Since PEG-derived hybridomas have an infinite life span and are continually secreting antibodies, it is essential that electroporation of B lymphocytes be performed in the presence of DNA sequences, which also contain so-called immortalization functions. That is, these DNA sequences confer on primary cells the ability to grow indefinitely in culture. Two such sequences exist in the early region 1A (E1A) of human adenovirus and the *myc* oncogene from the avian myelocytomatosis virus MC29.[51-53] Transfection of these genes into primary baby rat kidney cells or primary rat embryo fibroblasts resulted in the immortalization of these cells (without causing a tumorigenic conversion of the cells), thus allowing them to overcome cellular senescence or aging, a common problem when dealing with primary cells in culture.[52,53] Other genes adept at immortalizing cells are the large T oncogene of polyoma virus and the p53 tumor antigen gene.[54-57] Transfection and expression of these genes increase the proliferative potential of primary cells and facilitate their in vitro establishment. These "immortalizing" genes, in plasmid form, are readily available to the scientific community.

Another important consideration when deciding which DNA sequence should be transfected is the ability of the DNA to enhance or up-regulate the immortalization functions. Such sequences are entitled enhancers or transcriptional activators.[58-60] Enhancers belong to the set of eukaryotic regulatory elements that appears to increase the transcriptional efficiency of genes, independent of their position and orientation. Enhancer sequences can function upstream (i.e., toward the 5' end), within, or downstream (i.e., toward the 3' end) from eukaryotic genes. These activators can regulate transcription of genes over a considerable distance, perhaps even more than 10 kb. Several models have been proposed concerning the mechanism of action of enhancers, none of which have been proven definitively. These include the suggestion that enhancers function by providing a bi-directional entry site for RNA polymerase molecules or other factors associated with the transcription complex. This entry site model appears not to hold true for immunoglobulin enhancers. Alternatively, enhancers may alter the chromatin structure of a large region, which would allow the open conformation required for the initiation of transcription, or target promoters to active regions of the nuclear matrix. It seems more likely that the immunoglobulin enhancer binds particular proteins that can interact with those bound at the promoter sequences to create an efficient initiation complex. Enhancer elements for the immunoglobulins have been found for B cells. These enhancers function only in cells of lymphoid origin, and thus operate in a tissue-specific manner. Therefore, ligation of immunoglobulin enhancer sequences adjacent to sequences encoding immortalization functions would yield a molecule capable of specifically immortalizing B cells. In fact, such molecular constructs have been used for the production of transgenic mice, in which increased *myc* expression resulted from its close association with an immunoglobulin heavy-chain enhancer. The transgenic mice primarily developed an overproliferation of the pre-B-cell compartment, resulting in pre-B-cell lymphomas,[61] demonstrating that increased *myc* expression favored proliferation over maturation. The generation of another set of transgenic mice carrying a similar construct resulted in mice with tumors representing different stages of B-cell development.[62-64] In both instances, it appears that constitutive *myc* expression above a certain threshold promoted cell division at the expense of differentiation. It should be mentioned that transgenic mice carrying *myc* constructs lacking the immunoglobulin enhancers did not develop any tumors.[64] This is indicative of the efficacy and specificity of the constructs containing the enhancer elements. Therefore, such molecular constructs would seem to be good starting points for generating electrodomas, because the tissue specificity of the enhancer and the immortalizing capability of the *myc* oncogene have been well documented.

As mentioned previously, increased electroporation efficiency is critically dependent on the metabolic state and health of the cells. More specifically, the cells should be in the log phase of their growth curve, since quiescent cells are poor candidates for efficient uptake of DNA. When peripheral blood cells are collected, they are in a resting state. These cells must be activated in vitro or in vivo (i.e., regional draining lymph nodes, spleen cells) prior to electroporation. Many protocols are available in which B cells in particular are activated in vitro. These include stimulation with polyclonal activators, such as the mitogens phytohemaglutinin (PHA) and bacterial lipopolysaccharides (LPS), immunoglobulin-specific antibodies, or with antigen in the case of in vitro immunization. Regardless of the method chosen, it is essential to activate B cells before the electroporation procedure is performed.

II. PRELIMINARY EXPERIMENTS

Before describing the electroporation of mammalian cells, this chapter will describe a series of simple, yet critical, preliminary experiments that must be performed in order to optimize electroporation conditions and determine optimal culture selection conditions. Note that the basic molecular biology techniques (i.e., growth and transformation of bacterial cells, plasmid isolation and purification, restriction endonuclease digestion of DNA) will not be detailed here, because they are beyond the intended scope of this book. The reader is therefore urged to perform these techniques as presented in a methods manual, such as *Molecular Cloning: A Laboratory Manual*," by T. Maniatis, E. F. Frisch, and J. Sambrook, Cold Spring Harbor Laboratory (1982).

A. Activation of Peripheral Blood Cells

Activated or stimulated B cells should be used to obtain high-efficiency electroporation, as discussed earlier. If prior stimulation of B cells has been achieved, for example by in vivo activation or in vitro immunization procedures, the investigator can proceed to the determination of the appropriate drug concentration for selection conditions using these cells without further manipulations. In many cases, however, quiescent cells will be obtained from peripheral blood, in which case in vitro activation must be performed.

METHOD I: In Vitro Activation of Peripheral Blood Cells

1. Collect human peripheral blood into heparinized tubes.
2. Dilute blood 1:2 with complete culture medium (i.e., RPMI 1640 supplemented with 10% fetal calf serum, 2 mM L-glutamine, and 50 units/ml of a penicillin-streptomycin mixture).
3. Overlay 2 volumes of diluted blood onto 1 volume of Ficoll in sterile conical centrifuge tubes.
4. Centrifuge at 2000 rpm (500× g) for 20 min at room temperature.
5. Transfer white interfaces (containing peripheral blood lymphocytes) into another conical centrifuge tube and fill tube with medium.
6. Wash cells 2× (500× g per wash; 10 min per wash) at room temperature.

7. Resuspend cells in 4–5 ml of medium. Determine total cells obtained and their viability using trypan blue dye exclusion. Remember, approximately 1–2×10^6 cells are generally recovered from 1 ml blood.
8. Activate cells at a concentration of 1×10^6 cells/ml in complete culture medium. Add the mitogens PHA or LPS to final concentrations of 1 μg/ml or 1–10 μg/ml, respectively.
9. Culture for 3–5 days in a 37°C humidified atmosphere with 5% CO_2.
10. Harvest cells by gently scraping the cells from the flask using a rubber policeman, since activated lymphocytes generally attach to the culture surface. Transfer cells to a conical centrifuge tube.
11. Centrifuge at 2000 rpm (500 g) for 10 min at room temperature.
12. Wash cells 2× with 20 ml of complete medium (500 g per wash; 10 min per wash) at room temperature.
13. Resuspend cells in 2–10 ml of complete medium, depending on the starting number of cells.
14. Count cells and determine viability. The activated cells are now ready for future use.

If immune organs (spleen, tonsil, lymph node) are used as the source of activated cells, the following procedure is recommended. All dissections and cell preparations should be performed at room temperature.

METHOD II: *In Vitro Activation of Cells from Immune Organs*

1. Human tissue material should be received as soon as possible after surgery.
2. Transfer tissue to a culture dish containing 70% ethanol for 10–30 s.
3. Transfer tissue to a second culture dish. Immerse and rinse the tissue in a 5% penicillin-streptomycin mixture in serum-free culture medium.
4. Trim away excess tissue and capsular components with scissors and/or scalpel, using sterile technique.
5. Transfer needed material to another culture dish containing serum-free culture medium and tease apart the tissue with nugent forceps to obtain a single-cell suspension.
6. Transfer entire contents to a conical centrifuge tube and allow the large tissue aggregates to settle (approximately 5–10 min).
7. Transfer the cells in suspension to another centrifuge tube and wash cells 2× with complete culture medium (500 g per wash; 10 min per wash).
8. Resuspend cells in 2–10 ml of complete medium, depending on the size of the tissue specimen.
9. Count cells and determine viability.
10. Adjust cell concentration to 5×10^6 cells/ml.
11. Culture cells overnight in a 37°C humidified atmosphere with 5% CO_2.
12. Repeat steps 10–14 from Method I to harvest cells and prepare them for use.

B. Determination of Optimal Drug Concentration

The appropriate drug concentration for selection purposes should be determined by titration of the drug against the cells to be used for the electroporation experiments. If the investigator has decided on neo selection, geneticin (or G418) must be prepared as a stock solution. Dissolve G418 (Gibco) in culture medium *without* serum to approximately

40 mg/ml (actual drug concentration; note that G418 represents approximately 40–50% of the total weight of the drug preparation). The solution is then sterile filtered (0.2μ) and stored at 4°C. Alternatively, a stock solution of G418 may be prepared in 100 mM HEPES (N-2-hydroxyethylpiperazine-N′-2′-ethanesulphonic acid), pH 7.3, sterile filtered as above, and stored at −20°C. Addition of the drug should not alter the pH of the culture medium. Many investigators purchase a large amount of a single drug lot, because different lots can have different potencies. Thus, standardized selection conditions can be maintained.

If gpt selection is chosen, the investigator must remove guanine from the culture medium. It is therefore necessary to use dialyzed fetal calf serum and a medium that does not contain guanine (i.e., Dulbecco's modified Eagle's medium, Ham's F12 medium, RPMI 1640).

METHOD III: Titration of G418

1. Resuspend activated B cells to a concentration of 5×10^4 cells/ml in complete culture medium.
2. Aliquot 200 μl of cell suspension into multiple wells in a 96-well microtiter plate (1×10^4 cells/well).
3. To titrate the drug, add G418 to final concentrations ranging from 0–1200 μg/ml. Mix components in each well. Be sure to use 5–10 wells per drug concentration to accurately assess the effects of the drug.
4. Incubate in a 37°C humidified atmosphere with 5% CO_2 for 10–14 days.
5. Replace media, including G418, every 2–4 days, if necessary (i.e., if noticeable changes in the media are observed).
6. At the end of the incubation period, cell viability can be determined in the usual manner (i.e., trypan blue dye exclusion). Alternatively, supernatants can be aspirated and the cells washed with phosphate-buffered saline and then stained with 0.5% methylene blue in 50% methanol for 20 min.

The percentage of cell survival in the presence of each concentration of G418 versus the percentage of survival in the absence of any drug is calculated. A dose-response curve can then be obtained by plotting the percentage of survival versus the concentration of G418. Determine the minimal G418 concentration that kills 100% of the cells after 10–14 days. To minimize background, use a drug concentration that is slightly higher (by 25–50%) than the minimal G418 concentration determined for selection of the electrodomas. Typically, a drug concentration in the range of 400–800 μg/ml is used for mammalian cells. It should also be noted that: (1) cells will continue to divide for 1–2 generations even in the presence of lethal doses of G418, so the effects of G418 take several days to become apparent; and (2) G418 is most effective against dividing cells.

METHOD IV: Titration of Mycophenolic Acid

Mycophenolic acid can be prepared as a 50 mg/ml stock solution either in 0.1 N NaOH, neutralized with 0.1 M HCl, or in absolute ethanol. Sterile 1000× stock solutions of the following should also be prepared: xanthine (250 mg/ml), hypoxanthine (15 mg/ml), thymidine (10 mg/ml), aminopterin (2 mg/ml; light-sensitive), and L-glutamine (150 mg/ml). Complete culture medium consists of Dulbecco's modified Eagle's medium or RPMI 1640, 10% dialyzed fetal calf serum, 250 μg/ml xanthine, 15 μg/ml hypoxanthine (or 25 μg/ml adenine), 10 μg/ml thymidine, 2 μg/ml aminopterin, and 150 μg/ml

L-glutamine. Since the optimal lethal concentration of mycophenolic acid differs for different cells, the drug should be titrated in the *presence* and *absence* of guanine. Therefore complete culture medium, as described previously, should also contain 25 μg/ml guanine. This medium will be used *only* for determining the optimal drug concentration.

1. Resuspend activated B cells in complete culture medium (with or without guanine) to a concentration of 5×10^4 cells/ml.
2. Aliquot 200 μl of each cell suspension into multiple wells in a 96-well microtiter plate (1×10^4 cells/well).
3. Add mycophenolic acid to a final concentration of 0–100 μg/ml. Mix components in each well. Be sure to use 5–10 wells per drug concentration to obtain a better determination of the optimal concentration.
4. Incubate in a 37°C humidified atmosphere with 5% CO_2 for 7–14 days.
5. Replace medium, including all supplements, every 3 days.
6. Cell viability can then be determined as described earlier (refer to step 6 in Method III). A dose-response curve can be generated, from which the optimal lethal concentration of mycophenolic acid is established. A typical effective range of mycophenolic acid is 1–25 μg/ml. Mycophenolic acid alone is sufficient to prevent extensive growth of most cells. However, the addition of aminopterin will ensure complete inhibition of de novo synthesis of all purines. The presence of hypoxanthine (or adenine) and guanine in the culture medium overcomes this inhibition and allows the cells to synthesize purines via the salvage pathway.

C. Determining Optimal Electroporation Conditions

As mentioned earlier, successful electroporation is critically dependent on both biological variables and the electric field generated during gene transfer.[27] Most important, the electric field strength must be independently and accurately determined for each different cell type. Since only lymphocytes will be immortalized, matters are greatly simplified. As a general rule, increasing the applied voltage or the time constant causes increased cell damage and death until, at approximately 20–60% cell viability, maximum transfection occurs.[27,40] Further increases in the voltage or time constant produce higher levels of cell death and dramatic decreases in the transfection efficiency. Thus, a quick, though perhaps inaccurate, method of estimating the appropriate electroporation voltage is to determine the voltage at which approximately 50% cell viability occurs.

Another factor critical in determining the optimal electric field strength is the resistance of the electroporation buffer. Altering the buffer or the volume in the electroporation chamber affects the field strength. Decreasing the resistance (by increasing the ionic concentration of the buffer or the volume) results in shorter pulse times and smaller pore sizes, which could adversely affect transfection efficiency. It is relatively simple for the investigator to arrive at an optimal electric field strength for all experiments involving the same cell type by comparing cell viability and recovery using different field strengths and volumes.

METHOD V: Determining Optimal Field Strengths

1. Harvest activated cells from tissue culture flasks as described earlier (refer to step 10 in Method I).
2. Centrifuge at 2000 rpm ($500 \times g$) for 10 min at room temperature.

3. Wash cells 2× with culture medium *without* serum (500 × g per wash; 10 min per wash).
4. Resuspend cells to concentrations of approximately 1×10^7 cells/ml and 2×10^7 cells/ml in serum-free medium (or buffered saline solution) at room temperature.
5. Place 5×10^6 cells in the electroporation chamber.
6. Incubate cells at room temperature for 10 min.
7. Adjust the electroporation apparatus to deliver field strengths of 500–1000 V/cm (a typical range for lymphocytes) and a capacitance between 750–1000 μF. These settings are good starting points, and obviously depend on the electronic specifications of the instrument being used.
8. Place the electroporation chamber in the electroporation apparatus, connect the electrical leads, and apply the voltage according to manufacturer's instructions. Record the actual voltage delivered to the chamber, the time constant (in the millisecond range), and the actual capacitance.
9. After delivering the pulse, keep the cells at room temperature for 10 min.
10. Transfer the contents of the chamber to tissue culture flasks or dishes containing 10 ml of complete culture medium. Rinse the chamber with the same medium to recover all the cells and add to the flask or dish. Note that a significant amount of white cellular debris is visible in the chamber after the pulse is delivered.
11. Incubate in a 37°C humidified atmosphere with 5% CO_2.
12. Cell viability can be determined by trypan blue dye exclusion after 12–24 hr in culture. This allows the cells enough time for membrane resealing and recovery from the electroporation procedure.
13. Repeat steps 5–12, but: (1) vary the volume in the chamber by 50 μl increments, keeping the cell number constant; and (2) vary the voltage applied, in 50 V increments.
14. A control sample (i.e., not exposed to an electric pulse) should be treated in the same manner. This will enable the investigator to determine the total recovery of cells after electroporation.

It will be observed that cell viability and cell recovery decrease as the electric field strength increases. Higher voltages or longer pulse times should be used if no cell death is observed. Conversely, if most of the cells are killed, try lower voltages or shorter time constants. The field strength and volume chosen for future experiments should be the one that yields approximately 50% cell viability and reasonable recoveries.

Simply determining the voltage at which approximately 50% cell viability occurs is a quick method of arriving at an optimal electric field strength.[26,27] This method allows the investigator to become familiar with the electroporation apparatus and the required manipulations of the cells, but it can be inaccurate. The method of choice is to electroporate cells in the presence of varying concentrations of a marker plasmid, culture them in selective medium, and calculate transfection efficiency. In order to accomplish this, the linearized plasmid must be available. Typically, 1–20 μg/ml of linearized plasmid DNA is used for stable transfection.

METHOD VI: Determining Optimal Field Strengths

1. Harvest and resuspend cells as in steps 1–4, Method I.
2. Place 5×10^6 cells in the electroporation chamber.
3. Add linearized plasmid DNA (in sterile 10 mM Tris-HCl, pH 7.4, 1 mM EDTA) to a final concentration of 1–20 μg/ml. Mix gently.

4. Incubate the mixture for 10 min at room temperature.
5. Repeat steps 7–11 from Method I.
6. After 48 hr in culture, replace culture medium with selective medium containing the appropriate concentration of the antibiotic, as determined earlier.
7. As detailed in Method I, the volume and voltage applied should be varied in order to arrive at optimal electroporation conditions.
8. Several control samples are required: (1) cells not exposed to an electric pulse or plasmid DNA; (2) cells incubated with plasmid DNA but not exposed to an electric pulse; and (3) samples #1 and 2 incubated in selective (with the drug) or nonselective (without any drugs) medium.
9. The transfection efficiency can be determined after 10–14 days in culture by counting the number of antibiotic-resistant colonies.

The maximum transfection efficiency will be represented by a reproducibly sharp peak.[26] Therefore, the voltage and volume yielding the highest transfection efficiency should then be used for the generation of electrodomas. In the control samples: (1) cells not exposed to DNA or an electric pulse will thrive in nonselective media, but will die in selective media; (2) a few drug-resistant colonies may be observed in control cells incubated with DNA, not exposed to a pulse, and cultured in selective media, because DNA can simply adsorb to and be spontaneously transferred across the cellular membrane, albeit at an extremely low efficiency; and (3) cells incubated with DNA, not exposed to a pulse, and cultured in nonselective media will also thrive.

III. HYBRIDOMA GENERATION

After all variables have been examined and optimal electroporation conditions determined experimentally, it is time to prepare for the generation of electrodomas via the recombinant DNA technology route. One major stumbling block is the decision concerning which DNA sequences are to be used for efficient immortalization of B lymphocytes. Plasmids containing an immortalizing oncogene such as *myc* seem to be ideal. Unless precautions are taken to obtain a "pure" population of B cells (i.e., by sheep erythrocyte rosetting), however, the investigator will undoubtedly perform electroporation on a mixed population of cells, resulting in the immortalization of some unwanted cell types. This can be overcome by using a plasmid that contains not only an immortalizing gene, but also a DNA sequence that is active only in B lymphocytes. Such sequences as the immunoglobulin enhancers can be transfected into many cell types, yet are transcriptionally active only in B cells. These plasmids, with *myc* coupled to the immunoglobulin enhancer for the mu heavy chain or the kappa light chain, are available from the original investigators.[64] The reader should therefore contact these scientists to obtain the needed plasmids.

Another important consideration is that cotransfection of a marker plasmid (i.e., the plasmid conferring drug resistance) and the "immortalizing" plasmid is necessary. Generally, a 5:1 or 10:1 molar ratio of the plasmid containing the gene(s) of interest to the plasmid containing the selective marker is used. It has been reported that cotransfection of two separate plasmids by electroporation yielded cotransfection efficiencies in the range of 23–100%.[28,40] Therefore, transfected cells that grow in selective media have a high probability of also expressing the gene(s) of interest. This demonstrates that it is not essential to construct a plasmid containing both a drug-resistance gene and the gene of interest, thus simplifying the manipulations the investigator must perform.

A brief description of the transfection of B lymphocytes by electroporation follows. Immediately thereafter is a detailed protocol for electroporation, resulting in the generation of recombinant cells secreting immunoglobulin. A suspension of actively growing cells is placed into an electroporation chamber, and DNA (in the form of linearized plasmids) is added. Following a brief incubation period, the chamber is connected to a power supply, and the cells are subjected to a high-voltage DC electric pulse of defined magnitude and length. After a brief recovery period, the cells are placed in normal growth medium. After 24–48 hr, the cells are transferred to selective medium and allowed to proliferate. Within 10–14 days after electroporation, recombinant cells can be recovered, expanded, and analyzed for antibody production.

METHOD VII: Generation of Immunoglobulin-Secreting Recombinant Cells through Electroporation

1. Harvest activated B lymphocytes by scraping the tissue culture flask with a rubber policeman and transferring the cells to centrifuge tubes.
2. Centrifuge the cells at $500 \times g$ for 20 min at room temperature.
3. Wash the cells $2\times$ with 10–20 ml of room temperature serum-free culture media ($500 \times g$, 10 min per wash, at room temperature).
4. Resuspend cells in room temperature serum-free media. Determine the total cell number and cell viability.
5. Adjust cell concentration to $1–2 \times 10^7$ viable cells/ml in serum-free media. If necessary, centrifuge cells as above before adjusting cell concentration.
6. Place $3–5 \times 10^6$ viable cells in the electroporation chamber.
7. Add sterile, linearized plasmid DNAs (either in sterile 10 mM Tris HCl, pH 7.4, 1 mM EDTA, or sterile H_2O) to the cell suspension and incubate at room temperature for 10 min. The plasmids should be linearized by restriction enzyme digestion at a site distant from the marker or "immortalizing" gene. Use the previously determined optimal amounts of DNA in a molar ratio of 5:1 or 10:1 of the plasmid containing the "immortalizing" gene to the plasmid containing the marker (i.e., drug-resistance) gene, respectively. Be sure to include the appropriate controls, as discussed earlier.
8. Place the electroporation chamber in the holder, connect the electrical leads, and apply a pulse at the previously optimized voltage and capacitance settings.
9. Remove the chamber from the electroporation unit and incubate the pulsed sample for 10 min at room temperature.
10. Transfer contents of the electroporation chamber to a T75 tissue culture flask containing 10 ml of complete culture medium. Wash the chamber with medium and combine contents in the flask.
11. Incubate for 48 hr in a 37°C humidified atmosphere with 5% CO_2.
12. Harvest cells from each flask into separate 15-ml conical centrifuge tubes and collect cells by centrifugation at $500 \times g$ for 10 min at room temperature.
13. Resuspend each pellet in 10 ml of selective culture medium and perform a cell count. Aliquot 200 μl containing $10^2–10^4$ viable cells per well in a 96-well microtiter plate.
14. Incubate in a 37°C humidified atmosphere with 5% CO_2, replacing the media and antibiotic every 2–4 days.
15. After 10–14 days in culture, drug-resistant colonies should become visibly apparent.
16. Supernatants from the wells can be screened for immunoglobulin production using one of several of the commonly used detection methods: radioimmunoassay (RIA), enzyme-linked immunosorbent assay (ELISA), or fluorescence immunoassay (FIA). Typically, a nonisotopic method such as ELISA or FIA is

chosen.[65] The details for these techniques will not be given here, as they can be found in many laboratory manuals. Those cells secreting high levels of immunoglobulins can then be expanded in culture, and the nature of the secreted antibody can be determined by a variety of biochemical and immunological methods.

Points to Consider

1. The B lymphocytes used for electroporation must be previously activated so they are in the log phase of their growth curve. Actively proliferating cells appear to have an increased ability to take up and incorporate DNA.

2. All manipulations should be performed at room temperature, because this has also been shown to increase transfection efficiency.

3. The solution selected to wash the cells in prior to electroporation (i.e., buffered saline solution, serum-free medium, or complete culture medium) is left to the discretion of the investigator. However, the solution chosen must not adversely affect the cells in any manner.

4. It has been reported that electroporation efficiency was not affected by the presence of serum. However, because the composition of serum is unknown, it is possible that enzymes capable of degrading, or agents capable of denaturing, linearized plasmid DNA are present. Thus it may be best to electroporate cells in serum-free culture medium or buffered saline solutions.

5. If gpt selection is chosen, be sure to use dialyzed calf serum, media that is devoid of guanine for culturing, and the optimal concentration of mycophenolic acid. If selection is based on the neo gene, be sure to use G418 at a concentration slightly higher than the experimentally determined optimal G418 concentration.

6. If the transfected cells can be grown in serum-free media, characterization of the secreted immunoglobulins will be greatly simplified.

7. Each permanent transfection usually requires $3-5 \times 10^6$ cells to yield a reasonable number of transfectants.

8. Cells are plated at a low concentration in selective medium in order to minimize cross-contamination between electrodomas. This simplifies the cloning of a single electrodoma, and in turn, the purification of the secreted antibody.

IV. CONCLUSIONS

A variety of biochemical methods has been developed to achieve the transfer of genes into eukaryotic cells as a means for studying gene regulation and expression. Some cell types, however, are particularly refractory to transfection mediated by chemical treatment or dependent on biological vectors. For example, the transfection efficiency of hematopoietic cells is generally low when chemical methods are used. However, physical methods have also been devised that lead to enhanced DNA uptake across cellular membranes. Briefly exposing eukaryotic cells to a high-voltage DC electric pulse results in the transient breakdown of localized areas in the cellular membrane, thereby increasing the permeability of the cells to exogenous macromolecules. This technique of electroporation is rapidly becoming the method of choice for transfecting lymphocytes, as well as many other cell types of both eukaryotic and prokaryotic origin. Electroporation efficiencies have been reported to approach 1%, which, on the average, is approximately 2-3 orders of magnitude more efficient than transfection by chemical methods,[26,35,40,41] and 3-4 orders of magnitude more efficient than traditional PEG fusions. Electroporation is relatively inefficient, however, when compared to microinjection of DNA[19] or retrovirus-mediated gene transfer (10–50% frequency of transformation). A major advantage of electroporation is that it can be performed in any laboratory without specially acquired technical skills, which are required for microinjection, for example.

Efficient transfection is dependent upon the optimization of both electrical and biological parameters for each individual cell type.[22–24,26–28,31,32,34,35,66] The physical parameters include the electric field strength, pulse decay time, and the temperature at which electroporation is performed. The biological variables include the cell type, condition of the cells, electroporation medium, and the DNA concentration and topology. Once optimal conditions have been determined for a particular cell type, electroporation is straightforward. Detailed protocols for the experimental determination of optimal electroporation conditions have been included in this chapter.

The ability of mammalian cells to express foreign genes was well established before the advent of DNA-mediated gene transfer techniques. Studies of interspecies cell hybrids and genes transferred from isolated metaphase chromosomes clearly indicated that mammalian cells could express genes from other species. Expression is not limited only to eukaryotic genes. Prokaryotic genes can also be expressed, but they must be under the control of eukaryotic transcription signals. The expression of prokaryotic genes has proven to be an invaluable tool in selecting transformants. Two of the most commonly used dominant selectable genetic markers, or those that can be expressed in wild-type mammalian cells, have been discussed in detail.

The generation of immortalized, immunoglobulin-secreting lymphocytes via recombinant DNA technology is dependent upon the proper selection of the DNA sequences which are to be transfected. Many reports have appeared in the literature describing genes that encode for immortalization functions. It is essential, however, that the investigator also choose genes or DNA sequences whose expression is limited to cells in the hematopoietic lineage, and more specifically to B lymphocytes. For this reason, sequences encoding the immunoglobulin enhancers seem to be ideal candidates for targeting B cells, as they do not function in nonlymphoid cells, and are largely responsible for the tissue-specific expression of immunoglobulin genes. Evidence for this targeting of B cells was provided by coupling the *myc* oncogene, which is active in a variety of tissues, to the enhancer for the immunoglobulin heavy chain, and producing transgenic mice with this molecular construct.[61–64] More than 95% of the transgenic mice developed tumors of the B-cell lineage. Thus the specificity of the diseases apparently was derived from control imparted by the enhancer. Therefore, constructs such as these, which contain sequences for immortalization and tissue-specificity, target the action of these genes to particular cell types—in this case B cells.

In summary, recombinant DNA technology has entered the realm of immunology and offers an alternative to the more traditional methods of generating immortalized, immunoglobulin-secreting lymphocytes. The study of the molecular mechanisms governing gene regulation and expression has been made easier with the advent of various transfection techniques. Transfection by electroporation is a simple, useful, highly reproducible, and efficient method for introducing foreign genes into cells. It holds promise for a more efficient means of producing a series of new human monoclonal antibodies for diagnostic as well as therapeutic uses. Genetic engineering of lymphocytes derived from the immune organs of affected individuals could produce genetically stable, immunoglobulin-secreting, immortalized lymphocytes. These reactive antibodies could then potentially be used for immunotherapy, as they could be made patient-specific.[67]

REFERENCES

1. Kohler, G., and Milstein, C. 1975. Continuous cultures of fused cells secreting antibody of predefined specificity. Nature 256:495–497.

2. Graham, F. L., and Van der Eb, A. J. 1973. A new technique for the assay of infectivity of human adenovirus 5 DNA. Virology 52:456–467.

3. Robins, D. M., Ripley, S., Henderson, A. S., and Axel, R. 1981. Transforming DNA integrates into the host chromosome. Cell 23:29–39.

4. Wigler, M., Silverstein, S., Lee, L-S., Pellicer, A., Cheng, Y-C., and Axel, R. 1977. Transfer of purified herpes virus thymidine kinase gene to cultured mouse cells. Cell 11:223–232.

5. Oi, V. T., Morrison, S. L., Herzenberg, L. A., and Berg, P. 1983. Immunoglobulin gene expression in transformed lymphoid cells. Proc. Natl. Acad. Sci. USA 80:825–829.

6. Farber, F. E., Melnick, J. L., and Butel, J. S. 1975. Optimal conditions for uptake of exogenous DNA by Chinese hamster lung cells deficient in hypoxanthine guanine phosphoribosyltransferase. Biochim. Biophys. Acta 390:298–311.

7. McCutchan, J. H., and Pagano, J. S. 1968. Enhancement of the infectivity of simian virus 40 deoxyribonucleic acid with diethylaminoethyl-dextran. J. Natl. Cancer Inst. 41:351–357.

8. Miller, A. D., Jolly, D. J., Friedmann, T., and Verma, I. M. 1983. A transmissible retrovirus expressing human hypoxanthine phosphoribosyltransferase (HPRT): Gene transfer into cells obtained from humans deficient in HPRT. Proc. Natl. Acad. Sci. USA 80:4709–4713.

9. Mulligan, R. C., Howard, B. H., and Berg, P. 1979. Synthesis of rabbit β-globin in cultured monkey kidney cells following infection with a SV40 β-globin recombinant genome. Nature 277:108–114.

10. Fraley, R., Subramani, S., Berg, P., and Papahadjopoulos, D. 1980. Introduction of liposome-encapsulated SV40 DNA into cells. J. Biol. Chem. 255:10431–10435.

11. Schaefer-Ridder, M., Wang, Y., and Hofschneider, P. H. 1982. Liposomes as gene carriers: Efficient transformation of mouse L cells by thymidine kinase gene. Science 215:166–168.

12. Felgner, P. L., Gadek, T. R., Holm, M., Roman, R., Chan, H. W., Wenz, M., Northrop, J. P., Ringold, G. M., and Danielsen, M. 1987. Lipofection: A highly efficient, lipid-mediated DNA-transfection procedure. Proc. Natl. Acad. Sci. USA 84:7413–7417.

13. de Saint Vincent, B. R., Delbruck, S., Eckhart, W., Meinkoth, J., Vitto, L., and Wahl, G. 1981. The cloning and reintroduction into animal cells of a functional CAD gene, a dominant amplifiable genetic marker. Cell 27:267–277.

14. Schaffner, W. 1980. Direct transfer of cloned genes from bacteria to mammalian cells. Proc. Natl. Acad. Sci. USA 77:2163–2167.

15. Sugden, B., Marsh, K., and Yates, J. 1985. A vector that replicates as a plasmid and can be efficiently selected in B-lymphoblasts transformed by Epstein-Barr virus. Mol. Cell. Biol. 5:410–413.

16. Rassoulzadegan, M., Binetruy, B., and Cuzin, F. 1982. High frequency of gene transfer after fusion between bacteria and eukaryotic cells. Nature 295:257–259.

17. Sandri-Goldin, R. M., Goldin, A. L., Levine, M., and Glorioso, J. C. 1981. High-frequency transfer of cloned herpes simplex virus type 1 sequences to mammalian cells by protoplast fusion. Mol. Cell. Biol. 1:743–752.

18. Anderson, W. F., Killos, L., Sanders-Haigh, L., Kretschmer, P. J., and Diacumakos, E. G. 1980. Replication and expression of thymidine kinase and human globin genes microinjected into mouse fibroblasts. Proc. Natl. Acad. Sci. USA 77:5399–5403.

19. Capecchi, M. R. 1980. High efficiency transformation by direct microinjection of DNA into cultured mammalian cells. Cell 22:479–488.

20. Thomas, K. R., Folger, K. R., and Capecchi, M. R. 1986. High frequency targeting of genes to specific sites in the mammalian genome. Cell 44:419–428.

21. Zimmermann, U. 1982. Electric field-mediated fusion and related electrical phenomena. Biochim. Biophys. Acta 694:227–277.

22. Neumann, E., Schaefer-Ridder, M., Wang, Y., and Hofschneider, P. H. 1982. Gene transfer into mouse lyoma cells by electroporation in high electric fields. EMBO J. 1:841–845.

23. Wong, T-K., and Neumann, E. 1982. Electric field mediated gene transfer. Biochem. Biophys. Res. Commun. 107:584–587.

24. Potter, H., Weir, L., and Leder, P. 1984. Enhancer-dependent expression of human κ immunoglobulin genes introduced into mouse pre-B lymphocytes by electroporation. Proc. Natl. Acad. Sci. USA 81:7161–7165.

25. Zimmermann, U., and Vienken, J. 1982. Electric field-induced cell-to-cell fusion. J. Memb. Biol. 67:165–182.

26. Chu, G., Hayakawa, H., and Berg, P. 1987. Electroporation for the efficient transfection of mammalian cells with DNA. Nucl. Acids Res. 15:1311–1326.
27. Knutson, J. C., and Yee, D. 1987. Electroporation: Parameters affecting transfer of DNA into mammalian cells. Anal. Biochem. 164:44–52.
28. Toneguzzo, F., Hayday, A. C., and Keating, A. 1986. Electric field-mediated DNA transfer: Transient and stable gene expression in human and mouse lymphoid cells. Mol. Cell. Biol. 6:703–706.
29. Knight, D. E. 1981. Rendering cells permeable by exposure to electric fields. Tech. Cell. Physiol. 113:1–20.
30. Knight, D. E., and Scrutton, M. C. 1986. Gaining access to the cytosol: The technique and some applications of electropermeabilization. Biochem. J. 234:497–506.
31. Toneguzzo, F., and Keating, A. 1986. Stable expression of selectable genes introduced into human hematopoietic stem cells by electric field-mediated DNA transfer. Proc. Natl. Acad. Sci. USA 83:3496–3499.
32. Tur-Kaspa, R., Teicher, L., Levine, B. J., Skoultchi, A. I., and Shafritz, D. A. 1986. Use of electroporation to introduce biologically active foreign genes into primary rat hepatocytes. Mol. Cell. Biol. 6:716–718.
33. Bradshaw, H. D., Jr., Parson, W. W., Sheffer, M., Lioubin, P. J., Mulvihill, E. R., and Gordon, M. P. 1987. Design, construction, and use of an electroporator for plant protoplasts and animal cells. Anal. Biochem. 166:342–348.
34. Hama-Inaba, H., Takahashi, M., Kasai, M., Shiomi, T., Ito, A., Hanaoka, F., and Sato, K. 1987. Optimum conditions for electric pulse-mediated gene transfer to mammalian cells in suspension. Cell Struct. Funct. 12:173–180.
35. Spandidos, D. A. 1987. Electric field-mediated gene transfer (electroporation) into mouse Friend and human K562 erythroleukemic cells. Gene Anal. Techn. 4:50–56.
36. Kinosita, K., Jr., and Tsong, T. Y. 1977. Voltage-induced pore formation and hemolysis of human erythrocytes. Biochim. Biophys. Acta 471:227–242.
37. Kinosita, K., Jr., and Tsong, T. Y. 1977. Formation and resealing of pores of controlled sizes in human erythrocyte membranes. Nature 268:438–441.
38. Stopper, H., Jones, H., and Zimmermann, U. 1987. Large scale transfection of mouse L-cells by electropermeabilization. Biochim. Biophys. Acta 900:38–44.
39. Bertling, W., Hunger-Bertling, K., and Cline, M. J. 1987. Intranuclear uptake and persistence of biologically active DNA after electroporation of mammalian cells. J. Biochem. Biophys. Meth. 14:223–232.
40. Boggs, S. S., Gregg, R. G., Borenstein, N., and Smithies, O. 1986. Efficient transformation and frequent single-site, single-copy insertion of DNA can be obtained in mouse erythroleukemia cells transformed by electroporation. Exp. Hematol. 14:988–994.
41. Jastreboff, M. M., Ito, E., Bertino, J. R., and Narayanan, R. 1987. Use of electroporation for high-molecular-weight DNA-mediated gene transfer. Exp. Cell Res. 171:513–517.
42. Miller, C. K., and Temin, H. M. 1983. High-efficiency ligation and recombination of DNA fragments by vertebrate cells. Science 220:606–609.
43. Michel, M. R., Elgizoli, M., Koblet, H., and Kempf, C. 1988. Diffusion loading conditions determine recovery of protein synthesis in electroporated P3X63Ag8 cells. Experientia 44:199–203.
44. Colbere-Garapin, F., Horodniceanu, F., Kourilsky, P., and Garapin, A-C. 1981. A new dominant hybrid selective marker for higher eukaryotic cells. J. Mol. Biol. 150:1–14.
45. Southern, P. J., and Berg, P. 1982. Transformation of mammalian cells to antibiotic resistance with a bacterial gene under control of the SV40 early region promoter. J. Mol. Appl. Genet. 1:327–341.
46. Davies, J., and Jimenez, A. 1980. A new selective agent for eukaryotic cloning vectors. Am. J. Trop. Med. Hyg. 29:1089–1092.
47. Chen, C., and Okayama, H. 1987. High-efficiency transformation of mammalian cells by plasmid DNA. Mol. Cell. Biol. 7:2745–2752.
48. Mulligan, R. C., and Berg, P. 1980. Expression of a bacterial gene in mammalian cells. Science 209:1422–1427.

49. Mulligan, R. C., and Berg, P. 1981. Selection for animal cells that express the *Escherichia coli* gene coding for xanthine-guanine phosphoribosyltransferase. Proc. Natl. Acad. Sci. USA 78:2072–2076.
50. Franklin, T. J., and Cook, J. M. 1969. The inhibition of nucleic acid synthesis by mycophenolic acid. Biochem. J. 113:515–524.
51. Houweling, A., Van den Elsen, P. J., and Van der Eb, A. J. 1980. Partial transformation of primary rat cells by the leftmost 4.5% fragment of adenovirus 5 DNA. Virology 105:537–550.
52. Ruley, H. E. 1983. Adenovirus early region 1A enables viral and cellular transforming genes to transform primary cells in culture. Nature 304:602–606.
53. Land, H., Parada, L. F., and Weinberg, R. A. 1983. Tumorigenic conversion of primary embryo fibroblasts requires at least two cooperating oncogenes. Nature 304:596–602.
54. Rassoulzadegan, M., Cowie, A., Carr, A., Glaichenhaus, N., Kamen, R., and Cuzin, F. 1982. The roles of individual polyoma virus early proteins in oncogenic transformation. Nature 300:713–718.
55. Rassoulzadegan, M., Naghashfar, Z., Cowie, A., Carr, A., Grisoni, M., Kamen, R., and Cuzin, F. 1983. Expression of the large T protein of polyoma virus promotes the establishment in culture of "normal" rodent fibroblast cell lines. Proc. Natl. Acad. Sci. USA 80:4354–4358.
56. Jenkins, J. R., Rudge, K., and Currie, G. A. 1984. Cellular immortalization by a cDNA clone encoding the transformation-associated phosphoprotein p53. Nature 312:651–654.
57. Eliyahu, D., Raz, A., Gruss, P., Givol, D., and Oren, M. 1984. Participation of p53 cellular tumour antigen in transformation of normal embryonic cells. Nature 312:646–649.
58. Khoury, G., and Gruss, P. 1983. Enhancer elements. Cell 33:313–314.
59. Atchison, M. L., and Perry, R. P. 1986. Tandem kappa immunoglobulin promoters are equally active in the presence of the kappa enhancer: Implications for models of enhancer function. Cell 46:253–262.
60. Ptashne, M. 1988. How eukaryotic transcriptional activators work. Nature 335:683–689.
61. Schmidt, E. V., Pattengale, P. K., Weir, L., and Leder, P. 1988. Transgenic mice bearing the human c-*myc* gene activated by an immunoglobulin enhancer: A pre-B-cell lymphoma model. Proc. Natl. Acad. Sci. USA 85:6047–6051.
62. Langdon, W. Y., Harris, A. W., Cory, S., and Adams, J. M. 1986. The c-*myc* oncogene perturbs B lymphocyte development in Eμ-*myc* transgenic mice. Cell 47:11–18.
63. Alexander, W. S., Schrader, J. W., and Adams, J. M. 1987. Expression of the c-*myc* oncogene under control of an immunoglobulin enhancer in Eμ-*myc* transgenic mice. Mol. Cell. Biol. 7:1436–1444.
64. Adams, J. M., Harris, A. H., Pinkert, C. A., Corcoran, L. M., Alexander, W. S., Cory, S., Palmiter, R. D., and Brinster, R. L. 1985. The c-*myc* oncogene driven by immunoglobulin enhancers induces lymphoid malignancy in transgenic mice. Nature 318:533–538.
65. Gaffar, S. A., and Glassy, M. C. Applications of human monoclonal antibodies in non-isotopic immunoassays. In Reviews on Immunoassay Technology, Vol. I, S. B. Pal, ed., Macmillan Press, Basingstoke, United Kingdom, 1988.
66. Stopper, H., Zimmermann, U., and Neil, G. A. 1988. Increased efficiency of transfection of murine hybridoma cells with DNA by electropermeabilization. J. Immunol. Meth. 109:145–151.
67. Glassy, M. C., and Dillman, R. O. 1988. Molecular biotherapy with human monoclonal antibodies. Mol. Biother. 1:7–13.

CHAPTER 5

B-Cell Hybridoma Production by Avidin-Biotin Mediated Electrofusion

Mary K. Conrad and Mathew M. S. Lo

I. INTRODUCTION

Kohler and Milstein's method for the production of cell lines secreting specific antibodies directed to an antigen of interest has provided major advancements in the biomedical sciences and is widely used. The method is based on the random fusion of two cell types in the presence of polyethyleneglycol (PEG),[1,2] although viruses were used in earlier experiments.[3] Antibodies to a wide variety of antigens have been reported. Spleen cells from hyperimmunized mice are fused to azaguanine-resistant myeloma cells (which lack the enzyme hypoxanthine guanine ribosyltransferase, and therefore die in the presence of aminopterin). Hybrids are selected in a medium containing aminopterin supplemented with hypoxanthine and thymidine (HAT). Unfused myeloma cells are killed by the aminopterin, and unfused spleen cells die off after about 1 to 2 weeks in culture. These and all chemically induced fusions are random and may produce several hundred hybrid colonies, which have to be screened for the secretion of the desired antibodies. Usually several experiments have to be performed to obtain monoclonal antibodies of the required specificity and antigen binding affinity.

Recently, cell fusion induced by intense electric fields has received much attention.[4-6] Electrofusion is usually 10 to 100 times more efficient that PEG fusion.[7] Higher efficiencies of up to 10,000 times over PEG efficiencies have been claimed,[6] but are not commonly found by workers in this field. Short, intense electric pulses are thought to generate minute pores in cell membranes. Cells in contact at the time of the pulse, or shortly thereafter,[8] fuse during the process of spontaneous membrane repair. This process is dependent, in part, on the temperature to which the cells are subjected after exposure to the electric pulse.

A critical step in electrofusion therefore is promoting cell-to-cell contact at the time of electric field application. Cells lying in a low-intensity alternating field come into contact through a process called dielectrophoresis, a biophysical phenomenon resulting in linear chains of cells.[6,9] Other means of promoting cell-cell adherence include the addition of pronase[10,11] or concanavalin A.[12] All of these methods for promoting membrane contact are random in nature, and therefore suffer from the same disadvantages as PEG fusions (and would be multiplied by 100-1000 if used in conjunction with electrofusion). Cell populations may be isolated in advance; however, such techniques are highly specialized, and attainable yield in the case of rare cell types may prove to be a limiting factor. Usually fewer than 1% of B cells in an immunized spleen secrete an antibody of interest.[13] Therefore, in order to benefit from the higher fusion efficiencies provided by electrofusion, investigators must devise methods promoting selective cell-to-cell contact.

II. SELECTIVE TARGETING USING ANTIGEN CONJUGATES

A targeting technique employing antigen coated myeloma cells was previously reported to promote fusion in the absence of any added chemical fusogen.[14] This method cannot be too efficient, because it depends solely on spontaneous fusion events promoted by the antigen coating. However, the technique points out a very useful basic concept that may be used to facilitate cell contact in a bulk heterogeneous cell suspension; B cells express convenient surface antibody receptors directed against a particular antigen. Furthermore, the well-characterized tight binding[15] between avidin and biotin[16,17] can be introduced into this system to achieve specific, preselected cell-to-cell contact prior to electrofusion. To accomplish this, myeloma cells are coated covalently with biotin (or avidin), and the antigen of interest is conjugated covalently to avidin (or biotin) (Fig. 5-1). Incubation of immunized spleen cells with the antigen-avidin conjugate and biotinylated myeloma cells,

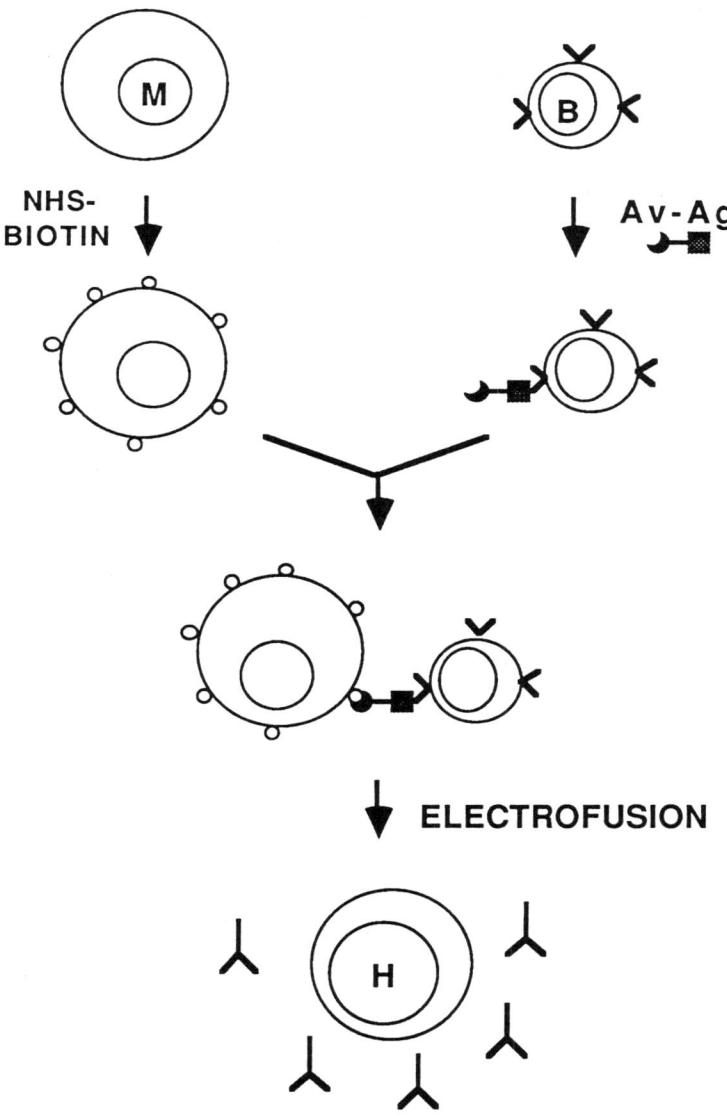

PRODUCING MONOCLONAL ANTIBODIES
BY TARGETING AND ELECTROFUSION

Fig. 5-1. Hybridoma production by bioselective electrofusion. Myeloma cells (M) are biotinylated with NHS-biotin. Cells prepared from immunized spleens are treated with an avidin-antigen conjugate (Av-Ag). B cells expressing cell surface antigen receptors (B) bind the antigen conjugate and become effectively coated with avidin. Myeloma and B-cell pairs are formed when the two cell types are mixed and incubated together. Cell pairs are fused in a high-voltage electric field. (This figure is taken from Conrad et al., 1987)

followed by electrofusion, produces relatively few hybrid colonies, a very high proportion of which secrete high-affinity antibodies directed specifically against the antigen used.[18]

A. Construction of Avidin-Antigen Conjugates

Conjugating avidin to antigen is a complex process that may be accomplished with a variety of chemical cross-linking procedures.[19] The essential steps consist of chemical activation of avidin (or antigen), reaction to antigen (or avidin), and purification of the conjugate by chromatographic separation based on size or charge differences. The reaction may be carried out in solution, or by immobilizing avidin onto iminobiotin-Sepharose.

1. Amino-Reactive Homobifunctional Cross-linker

The use of 1,5-difluoro-2,4-dinitrobenzene (DFDNB), a small homobifunctional cross-linker, in solution,[20-22] or on a solid phase[18] procedure, has been previously described. A large excess of DFDNB is used to activate avidin bound to iminobiotin-Sepharose. Proteins or peptides will now covalently conjugate to the activated, immobilized avidin through free amino groups.

Iminobiotin-Sepharose (usually 10–30 μmole iminobiotin/ml of packed Sepharose) is made by reacting ethylenediamine with cyanogen bromide activated Sepharose 4B-CL,[23] and then with the N-hydroxysuccinimide derivative of iminobiotin.[24] The complex is next blocked by reaction with Sanger's reagent (2,4-dinitro fluorobenzene). Following this, 70 μg of avidin is incubated with 0.1 ml of packed iminobiotin-Sepharose resin in 0.5 ml 0.1 M sodium borate buffer, pH 10.5, in a 1-ml column, at 23°C. for 1 hr. The resin is washed with 5 ml of sodium borate buffer to remove unreacted reagent. Next 0.1 nmole antigen is dissolved in the same buffer and added to the activated avidin resin, which is reacted for 18 hr at 23°C. Following this incubation, 2 ml of sodium borate buffer, pH 10.5, then 1 ml of buffer containing 1 mM lysine or glycine, are used to wash the column and block the unreacted DFDNB groups. Now the avidin-antigen conjugate may be eluted from the column with 0.1 M citrate buffer, pH 3.5. After addition of a carrier protein, such as bovine serum albumin (BSA), the conjugate is desalted on a Sephadex G-25 column, finally eluting with phosphate buffered saline (PBS). This solid-phase method of conjugation works well with large antigens, which are not significantly modified by altered basic residues.

Small peptides or acidic proteins, on the other hand, may become insoluble when conjugated in solid phase. These can be conjugated to avidin using DFDNB in solution. To do this, the molar ratio of the peptide to avidin must be reduced to less than 5 moles of peptide per mole of avidin, in order to prevent the formation of insoluble conjugate. This is a very efficient method; however, amino groups present on the antigen may react with DFDNB to form very large complexes, severely compromising the antigenicity of the protein.

2. Sulfhydryl-Reactive Heterobifunctional Cross-Linker

Another cross-linker that can be used is M-maleimidobenzoyl N-hydroxysuccinimide (MBS), which is heterobifunctional.[25,26] Cross-linking occurs through sulfhydryl groups on the antigen, and this is usually less detrimental to the antigenicity. In addition, because MBS reacts with lysine residues of avidin, the isoelectric point of the avidin-MBS complex is reduced to a more neutral pH value (beneficial, because native avidin is basic and binds acidic proteins, such as those on cell surfaces, nonspecifically). Avidin is reacted with a 5-molar excess of MBS for 1 hr at 23°C. Filtration on Sephadex-G25 removes free reagent. Absorbance at 282 nm allows calculation of the concentration of the complex.

Antigen is reduced with 10 mM dithiothreitol (DTT), then reacted with avidin-MBS for 20 hr. If desired, conjugates may be purified by gel filtration or ion-exchange chromatography, but in most cases they can be used without further purification.

3. Points to Consider on Coupling Antigens to Avidin

Haptens and carbohydrates can be chemically activated by reactions such as periodate oxidation to the N-hydroxysuccinimide derivative,[19] and then conjugated to avidin. This is usually more effective than using cross-linkers. Since the isoelectric point of reacted avidin is reduced, conjugates can be purified by ion-exchange chromatography. Avidin from egg white is more basic than strepavidin from *Streptomyces avidinii,* and is therefore preferred.

B. Construction of Biotin-Antigen Conjugates

Although constructing avidin-antigen conjugates is complicated, it is preferable to conjugation to biotin because of the relative ease of characterizing avidinylated conjugates and the greater efficiency of biotinylating myeloma cells. When the complexity of avidinylation cannot be overcome in some cases, however, biotin-antigen conjugates can easily be constructed for use with avidinylated myeloma cells.

1. Biotinylation

Derivatives of biotin or iminobiotin are used for biotinylation reactions. Commonly employed compounds have been N-hydroxysuccinimidyl (NHS) esters, which react with amines on proteins. NHS-biotin dissolved in DMSO (dimethylsulfoxide) or DMF (dimethylformamide) is added to the antigen in two- to tenfold molar excess. The reaction is carried out in PBS (pH 7.4), and excess biotin can be removed by gel filtration or dialysis.

Other reagents that can be used are maleimidyl and iodoacetyl derivatives of biotin. Free sulfhydryl groups and disulfhydryl bonds can be reduced prior to reaction with maleimide biotin. Biotinylation of cysteine residues on the antigen has fewer effects on the solubility and possibly the antigenicity of the molecule than does conjugation through amino groups using DFDNB or NHS.

2. Points to Consider on Biotinylation of Antigens

Proteins that are highly substituted with biotin become insoluble. Basic residues important for antigenicity may become biotinylated. In experiments to raise antibodies to unknown antigens, however, biotinylation may be the method of choice, because the reactions are simple, quantitative, and are less likely to result in antigenic modification than is avidinylation.

III. Testing the Conjugate

Before using a conjugate for selective biotargeting in a fusion experiment, one must ascertain that both the avidin (or biotin) and the antigen components of the conjugate are functional. Several approaches may be used toward this end, including radioimmunoassays (RIAs), enzyme-linked immunoabsorbent assays (EIAs), and fluorescent immunoassays. The latter will be described here. For the other methods the reader is referred to several references.[17,27] In general, an antibody raised to the antigen of interest coupled to a reporter group (e.g., radioisotope, fluorescent compound, or an enzyme) is employed in the assay. This may be an available and well-characterized monoclonal antibody, or serum from the immunized animal prepared for the actual fusion experiment.

A. Fluorescent Microsphere Assay

Brightly fluorescent microsphere beads are available commercially in different sizes and fluorescence wavelengths. These chemically activated microspheres may be rapidly and conveniently coated with proteins,[28,29] such as antibodies to the antigen of interest. Larger (0.7 or 0.9 μm) beads are obviously visible under low magnification with simple fluorescence illumination or even under phase contrast (since the fluorescence is so strong). Commercial beads (Covasphere MX; Duke Scientific Corp., CA) are resuspended by sonication for 30 s and reacted with up to 0.1 mg protein in 1 ml PBS for 2 hr at 23°C. Reacted beads are centrifuged in a microfuge at 10,000 $\times$ g for 10 min at 4°C, washed in 1 ml PBS containing 10 mg/ml BSA, and resuspended in 0.5 ml PBS. Prepared beads may be stored at 4°C for 3 months.

Assays consist of incubating various concentrations of the antigen-avidin conjugate, antibody-fluorescent microspheres (5–10 μl), and 0.5–1 $\times$ 10⁶ biotinylated myeloma cells (see below for protocol) at 23°C for 2 hr. Myeloma cells are washed in DME (in 10 $\times$ 75 mm glass test tubes, which minimizes cell loss), and resuspended in 100 μl DME. Aliquots (25 μl) are applied to a glass slide. The sample is sealed with a coverslip and nail polish. Samples are examined under fluorescence microscopy with low-intensity phase contrast illumination. Cells with more than 3 bound beads are scored positive. Fifty to 100 cells are scored for each sample. An example of this assay is shown in Fig. 5-2. Specific conjugate binding is evaluated by using unbiotinylated cells, omitting the conjugate from the assay, or replacing the conjugate with one that is unrelated to the antigen of interest.

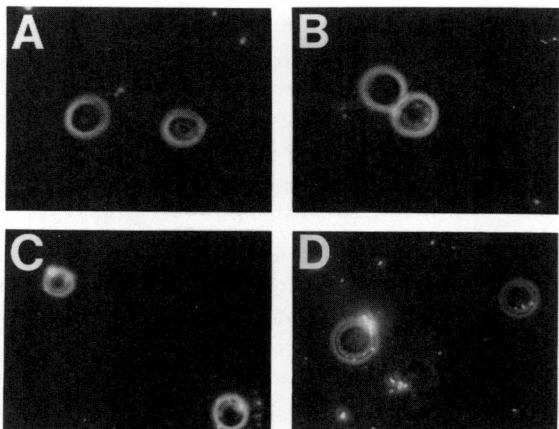

Fig. 5-2. Immunofluorescent assay of avidin-antigen binding to myeloma cells. Polystyrene microspheres, appearing as brightly fluorescent small beads in panels C and D, are coupled with a purified antihuman chorionic gonadotropin (hCG) antibody. These antibody-coated beads (mimicking B cells) are incubated with the avidin-hCG conjugate and biotinylated myeloma cells (C and D). Nonspecific binding is assessed by incubation with unbiotinylated myeloma cells (A) or with β-hCG alone (B). Low-intensity phase illumination is used to view cells. The 0.9-μm fluorescent microspheres appear as very bright spots.

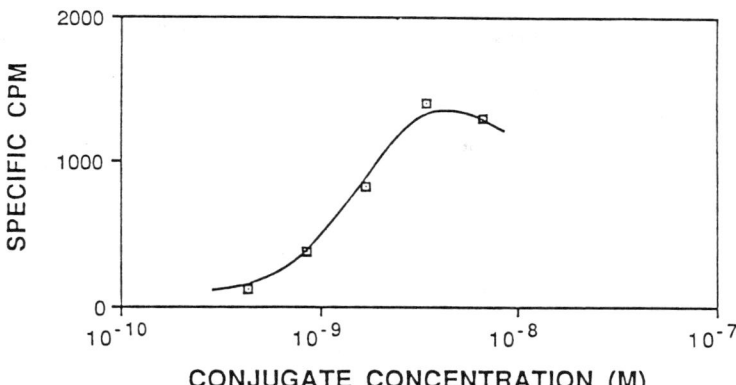

Fig. 5-3. Binding activity of β-hCG-Avidin. The same antibody used to coat the beads in Fig. 5-2 was iodinated and incubated with different concentrations of avidin-hCG conjugate and biotinylated myeloma cells, using the same conditions as in Fig. 5-2. Cells were harvested on a cell harvester to remove unbound ^{125}I-labeled antibody. Nonspecific binding was determined by using an irrelevant conjugate, or unbiotinylated myeloma cells (both gave less than 5% of total binding). The binding activity appears to be specific and dose-dependent.

B. Radioimmunoassay

The same assay may be reconfigured with ^{125}I-labeled antibodies in place of the fluorescent microspheres. This method, which is illustrated in Fig. 5-3, offers quantitation of conjugate binding activities, using the same conjugate preparation. Myeloma cells are harvested and washed with a cell harvester, which permits examination of larger sample numbers. Iodination and cell harvesting procedures may be found elsewhere.[27]

C. Points to Consider on Testing of Antigen Conjugate

Despite the effort required in testing antigen conjugates, confirmation of the antigenicity and avidin (or biotin) activity is important and cost-efficient. Conjugate binding properties may be predicted in most cases, and a suitable dilution determined in radioimmunoassays may be applied to the actual hybridoma fusion experiments. Similarly, enzyme-linked immunoassays may be employed, except biotinylated lysozyme or BSA is adsorbed onto a solid phase, which replaces the biotinylated myeloma cells (Fig. 5-4).

Qualitative properties of the conjugate may be judged from visual examination with the fluorescence assay. "Sticky" conjugates will aggregate the antibody-coated microspheres (i.e., more than 5–10 beads per aggregate), or bind nonspecifically to biotinylated cells. Conjugate binding should increase with increasing concentration of added conjugate. In some cases, binding activity will decline as the conjugate concentration is further increased. This may indicate the presence of competing unconjugated antigen or antigen covalently coupled to inactive avidin. Excess free avidin will aggregate biotinylated myeloma cells, but this will not interfere with the actual hybridoma fusion experiment where large numbers of myeloma cells are present.

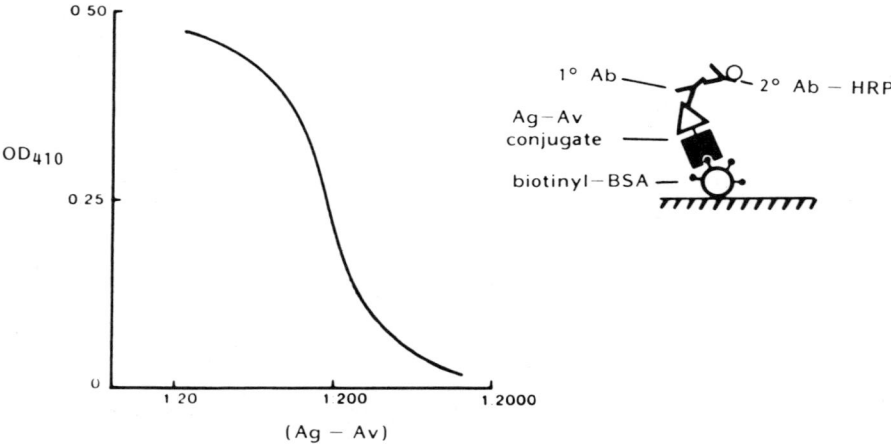

Fig. 5-4. Characterization of activity of avidin-antigen conjugate. Avidin conjugated to a peptide is first bound to plastic plates (Dynatek) coated with biotinylated bovine serum albumin (BSA). Plates are washed after incubation and a polyclonal antiserum raised against the peptide is added. After washing, a second antibody covalently attached to horseradish peroxidase (HRP) is added to the plate, which is subsequently washed and developed with substrate. The curve shows that the amount of bound conjugate, as measured by the optical density, can be detected to at least a 1:200 dilution of the original conjugate preparation.

IV. PREPARATION OF CELLS FOR FUSION

Before bioselectively targeted electrofusion using an antigen conjugate can occur, both myeloma and spleen cells must be prepared. Depending on the conjugate to be used, myeloma cells must be coated with either biotin or avidin. Spleen cells must be prepared in vivo by immunizing the animal. We have used mice and mouse myeloma cell lines for our fusions.

METHOD I: *Biotinylation of Myeloma Cells*

A number of myeloma cell lines are available for hybridoma production. Cell lines are screened for their ability to withstand chemical reaction with N-hydroxysuccinimidyl (NHS), as well as their ability to retain covalently conjugated biotin on the surface membrane. Four different cell lines tested (P3X63/Ag8.653, SP2/0, S194, FOX-NY) were all completely viable following biotinylation. However, the P3X63/Ag8.653 line retained 10% to 30% of the original surface biotin after overnight incubation, whereas the three other lines lost virtually all of the surface biotin. Therefore, we routinely used the P3X63/Ag8.653 line, hereafter referred to as P3.

Procedure

Log-phase cells are harvested and washed with Dulbecco's phosphate-buffered saline (PBS), containing no divalent cations. NHS-biotin is freshly prepared by dissolving NHS biotin in DMF (34 mg/ml), and diluting into PBS to a final concentration of 0.1 M. This solution is added immediately to the myeloma cell pellet, which is resuspended by gentle agitation of the tube. It is then incubated at room temperature for 15 min without further agitation. Serum (10 ml) is underlaid directly using a pipette, and the myeloma cell suspension is centrifuged through the serum at 200 $\times$ g for 8 minutes. (Less-expensive calf serum or fetal bovine serum may be used for this procedure, after heat inactivation at 56°C for 40 min.) Biotinylated myeloma cells are washed twice in Dulbecco's modified Eagle's medium (DME) containing deoxyribonuclease-1 (DNase). DME is used because it is biotin free. DNase is added to reduce cell aggregation caused by the release of intracellular components. DNase is removed at later stages because of possible intracellular introduction during electrofusion, which would be toxic.

Points to Consider on Biotinylation of Cells

The degree of biotinylation by this method may be tested by staining cells with fluoresceinated avidin. This test is recommended when this procedure is first employed. One very extensive procedure was used to sort approximately 2 million biotinylated P3 cells labeled with fluorescein-conjugated avidin, using a fluorescence-activated cell sorter (FACS) (Fig. 5-5). In our experiments, more than than 95% of the P3 cells were biotinylated. The degree of biotinylation varied by 2 log units of fluorescent intensity. Two cell fractions containing either heavily (fraction 5) or lightly (fraction 6) biotinylated cells

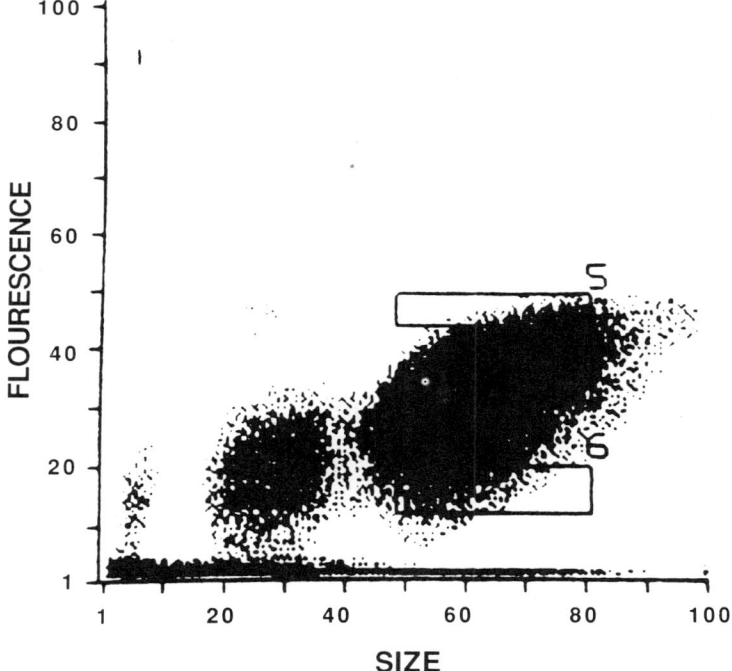

Fig. 5-5. Biotinylation of myeloma cells with NHS-biotin. Myeloma cells were biotinylated as described in the text. After washing, they are incubated with fluorescein-conjugated avidin. These were then sorted on a fluorescence-activated cell sorter (FACS), as shown. Cell size (axial coefficient) is measured as the cell crosses a laser beam. The fluorescent intensity ranges from 10–50 fluorescent units. Two populations of myeloma cells were collected, shown in the rectangles labeled 5 and 6. These were plated into soft agar to assess cell viability (determined by colony formation). The cloning efficiency of these two fractions did not differ from each other or from unlabeled myeloma cells plated at the same density.

were cloned in soft agar (according to the method of Civin and Banquerigo[30]). The viability of the two populations was scored by the number of resultant colonies, and was not found to be significantly different. Another more practical way of determining the degree of biotinylation is by visual examination with a fluorescence microscope after staining with fluorescein-conjugated avidin. In either case, cell viability can be assessed by trypan blue exclusion. Viability was found to be 98% or greater in our experiments.

METHOD II: Avidinylation of Myeloma Cells

When an antigen-biotin conjugate is to be used for fusion, myeloma cells must be avidinylated. This is a more complex procedure than biotinylation, but sometimes it is easier to make biotin-antigen conjugates, such as when dealing with large or complex proteins.

Procedure

Avidin is conjugated to the heterobifunctional cross-linker SPDP (N-succinimidyl 3-(2-pyridyldithio) propionate), which covalently attaches to myeloma cell surfaces. To form the conjugate, a 5-molar excess of SPDP is reacted with avidin for 1 hr at 23°C. Gel filtration on Sephadex-G25 with PBS is performed to remove free reagent. The absorbance of the conjugate at 253 nm and 282 nm allows quantitation of the reaction. Usually the ratio of avidin:SPDP is 4:1.

Myeloma cells are grown in medium containing 0.1 mM β-mercaptoethanol. About 10^7 cells growing in log phase are reacted with 1 μM avidin-SPDP at 23°C for 1 hr without agitation. Cells are centrifuged through a solution of equal volumes of DME and isosmotic sucrose. The extent of avidinylation can be determined by measuring the binding of biotinylated fluorescein-conjugated lysozyme. In our experiments, more than 80% of cells were successfully avidinylated; however, the fluorescence intensity was about tenfold lower than that of biotinylated myeloma cells stained with fluorescein-conjugated avidin.

The density of avidin binding to the myeloma cell surface can be greatly enhanced by prior reaction of the cells with reduced SPDP. This is made by dissolving 0.5 mg of dithiothreitol and 1 mg of SPDP in 30 μl of dimethylformamide, then diluting with 30 ml of PBS. Myeloma cells are incubated in this solution at 23°C for 15 min. Cells are centrifuged through sucrose/DME, then resuspended in PBS for incubation with 1 μM avidin-SPDP as described earlier.

Points to Consider on Avidinylation

We have found that the use of avidin to bridge biotinylated myeloma cells and biotinylated antigen is not reproducible, and often results in the formation of large aggregates of cells. It is therefore preferable to use avidinylated myeloma cells when using biotinylated antigen.

METHOD III: Preparing Cells for Fusion

Spleen cells used for fusions to produce monoclonal antibodies must first be primed in vivo with the antigen of interest, so that the animal's immune response will be stimulated, resulting in a subpopulation of mature B cells producing the desired antibodies. Following removal from the animal they must be carefully combined with myeloma cells prior to electrofusion.

Immunization of Mice

The animals usually used for hybridoma production are C57BL/6 or BALB/C mice. Antigen is usually injected intraperitoneally; if the antigen is precious, however, the final immunization can be a very small amount given by intravenous or intrasplenic injection.

Several approaches may be used to increase the antigenic response to foreign molecules. For most antigens, adsorption onto either bentonite or alumina is effective. Small molecules such as haptens or peptides are often conjugated to keyhole limpet hemocyanin (KLH) before adsorption. When this is done, it is often effective to immunize animals with KLH prior to injections with the KLH-antigen conjugate, because animals may more rapidly develop serum titers to the antigen. Adsorbed antigens may also be emulsified in Freund's adjuvant, although in our experience this results in increased problems with intraperitoneal adhesions.

The immunization schedule is quite flexible, and will depend upon the amount of antigen available and time considerations for the experiment. Compared with antibodies produced from acutely exposed animals, those from hyperimmunized animals are of greater variety and have higher affinities. An average experiment may include 3 or 4 injections of antigen 1–2 weeks apart. In our experiments, a very rare antigen was used, so animals were immunized twice only, intraperitoneally, 1 week apart, with microgram quantities. The final injection was given 3 days prior to fusion.

Isolation of Spleen Cells

Three days after the final immunization, animals are killed and washed thoroughly with an iodine-containing solution. Spleens are removed using sterile instruments, under a clean hood. The cells are dissociated into DME containing 50 μg/ml DNase, through a fine wire mesh using a sterile rubber policeman. The small amount of remaining fibrous tissue is discarded. Although not necessary, the red blood cells may be lysed if desired, using a freshly prepared 0.84% ammonium chloride solution.

Combining the Three Components for Fusion

The spleen cells are washed once with DME/DNase, then incubated in the same medium with the antigen-avidin (or antigen-biotin) conjugate for 30 min to 2 hr. The incubation is done without agitation at 4°C, to prevent capping and internalization of the conjugate.

Following incubation, the treated spleen cells are mixed with the treated myeloma cells in a 4:1 ratio, then centrifuged. Medium is aspirated, and the pellet is loosened and spun very lightly for 10 s at 50 $\times$ g. The cells are then incubated for 30 min at 23°C, or 37°C. DME is used to dilute the cells. An aliquot containing as many as 2×10^7 cells is spun through a sucrose underlayer at 200 $\times$ g for 6 min. The pellet is resuspended in 0.5 ml isosmotic sucrose (300–320 mOsm), and placed into the cell fusion chamber.

Points to Consider on Cell Preparation

Cells should be treated very gently at all stages of preparation. When resuspension of pellets is required, tapping or "finger flicking" should be used rather than vortexing or vigorous shaking. In addition, divalent cations cause nonspecific aggregation, and thus should be avoided. Isosmotic sucrose is used because of its low electrical conductance, in order to minimize heat generation during high voltage fusion.

METHOD IV: Cell Fusion

After completing the previously outlined steps, one is left with a suspension of cells in sucrose. In the suspension are the few B-cell–antigen–avidin–biotin–myeloma cell complexes of interest.

Electrofusion

A commercial electric pulse generator, monitored with a storage oscilloscope through a 100× reduction probe, is used for cell fusion.

Method

As many as 50 million cells, in 0.5 ml of sucrose, are introduced into the fusion chamber. The cells are subjected to 2 to 4 pulses, field intensity of 3 kV/cm, for 5 μs/pulse. The field polarity is switched between the pulses so that cells will not be transported in bulk to one or the other electrode.

Following fusion, cells are kept at 37°C for 30 min in DME. During this time cell membranes are undergoing the process of spontaneous repair of pores generated by the high voltage, a temperature-dependent process. Cells are then plated out into 96-well dishes in HAT medium.

Hybridoma growth is usually detected between 1 to 6 weeks following fusion. Hybridomas are screened for antibodies in the usual ways.[27]

V. CONCLUSIONS

We have used this method with a limited number of antigens. The first fusions were performed with ACE (angiotensin converting enzyme) conjugated to avidin. Three fusions were done, and all resultant hybrid colonies produced antibodies to ACE (a total of 48 clones). We obtained predominantly IgG antibodies, which was unexpected because we used an acute immunization regime (described earlier). We believe that the bioselection using antigen conjugate facilitated fusion of those B cells that are producing high-affinity, IgG-type antibodies, whereas more weakly binding antibody-producing spleen cells are less likely to be fused.

We feel the ACE fusions were particularly successful because ACE is a large glycoprotein (180,000 daltons) that is very resistant to modification by the conjugation process. Subsequent fusions with small peptides, conjugated with DFDNB, gave high yields, in the 10–50% range, but we did not have 100% success with these antigens. In addition, the affinities of these antibodies were about $K_D = 10^{-8}$ M, lower than the anti-ACE antibodies by ten- to twentyfold.

The preceding fusions were not done in parallel with PEG fusions, so a large experiment was performed using β-hCG (β-human chorionic gonadotropin) for the antigen. This was conjugated to avidin with MBS on a large scale, and was shown to be fully functional with respect to both the avidin component and the antigen component. The molecular weight of the conjugate on SDS gels showed a 1:1 stoichiometry (antigen:avidin). Spleens from 6–8 mice were pooled for each of four experiments. For each, six conditions, each with equal cell numbers, were used: for electrofusions, five concentrations of conjugate were used, and the remaining cells were fused with PEG, and without conjugate. Electrofusions resulted in 12.8% of growing hybrids producing anti-β-HCG antibodies, whereas the PEG fusions gave yields of only 2.7% (combined results of the four experiments). Because of the screening method used, we could only detect antibodies with affinities higher than $K_D = 10^{-10}$ M. When lower affinity antibodies were screened, the results were 7.3% for PEG and 39.4% for the electrofusions.

We therefore believe that high-voltage fusions offer advantages, especially in that the numbers of growing hybrid colonies are relatively small, making the screening process very cost-effective compared with PEG fusions. Because the high-voltage fusion process is very new and is still being developed, the future may reveal even more advantages over the well-established method of PEG fusion for the production of monoclonal antibodies.

REFERENCES

1. Galfre, G., Howe, S. C., Milstein, C., Butcher, G. W., and Howard, J. C. 1977. Antibodies to major histocompatibility antigens produced by hybrid cell lines. Nature 266:550–552.
2. Galfre, G., and Milstein, C. 1981. Production of monoclonal antibodies: Strategies and procedures. Methods Enzymol. 73:3–46.
3. Kohler, G., and Milstein, C. 1975. Continuous cultures of fused cells secreting antibody of predefined specificity. Nature 256:495–497.
4. Neumann, E., Gerisch, G., and Opatz, K. 1980. Cell fusion by high electric impulses applied to dictyostelium-discoideum. Naturwissenschaften 67:414–415.
5. Tsong, T. Y. 1983. Voltage modulation of membrane permeability and energy utilization in cells. Biosci. Rep. 3:487–505.
6. Zimmermann, U., Vienken, J., Halfmann, J., and Emeis, C. C. 1985. Electrofusion: A novel hybridization process. Adv. Biotech. Processes 4:79–150.
7. Karsten, U., Papsdorf, G., Roloff, G., Stolley, P., Abel, H., Walther, I., and Weiss, H. 1985. Monoclonal anti-cytokeratin antibody from a hybridoma clone generated by electrofusion. Eur. J. Cancer Clin. Oncol. 21:733–740.
8. Sowers, A. E. 1986. A long-lived fusogenic state is induced in erythrocyte ghosts by electric pulses. J. Cell Biol. 102:1358–1362.
9. Pohl, H. A. Dielectrophoresis, Cambridge University Press, Cambridge, 1978.
10. Vienken, J., and Zimmermann, U. 1982. Electric field-induced fusion: Electro-hydraulic procedure for production of heterokaryon cells in high yield. FEBS Lett. 137:11–13.
11. Bischoff, R., Eisert, R. M., Schedel, I., Vienken, J., and Zimmermann, U. 1982. Human hybridoma cells produced by electrofusion. FEBS Lett. 147:64–68.
12. Conrad, M. K., Lo, M. M. S., Tsong, T. Y., and Snyder, S. H. In Cell Fusion, A. Sowers, ed., Plenum Press, New York, 1987.
13. Golub, E. B. 1987. Somatic mutation: Diversity in regulation of the immune repertoire. Cell 48:723–724.
14. Bankert, R. B., DesSoye, D., and Power, L. 1980. Antigen-promoted cell fusion: Antigen-coated myeloma cells fused with antigen-reactive spleen cells, Transplant. Proc. 12:443–448.
15. Green, N. M. 1975. Avidin. Adv. Protein Chem. 29:85–133.
16. Godfrey, W., Doe, B., Wallace, E. F., Bredt, B., and Wofsy, L. 1981. Affinity targeting of membrane vesicles to cell surface. Exp. Cell Res. 135:137–145.
17. Wormmeester, J., Stiekema, F., and Groot, K. D. 1984. A simple method for immunoselective cell separation with avidin-biotin system. J. Immunol. Method 67:389–394.
18. Lo, M. M. S., Tsong, Y. T., Conrad, M. K., Strittmatter, S. M., Hester, L. H., and Snyder, S. H. 1984. Monoclonal antibody production by receptor-mediated electrically induced cell fusion. Nature 310:792–794.
19. Tijssen, P. Practice and Theory of Enzyme Immunoassays, Vol. 26. Elsevier, New York, 1985.
20. Goodfriend, T. L., Levine, L., and Fasman, G. D. 1964. Antibodies to bradykinin and angiotensin: A use of carbodimides in immunology. Science 144:1344–1346.
21. Tager, H. S. 1976. Coupling of peptides to albumin with difluorodinitrobenzane. Analyt. Biochem. 71:367–375.
22. Golds, E. E., and Braun, P. E. 1978. Protein association in basic protein conformation in the myelin membrane: The use of difluorodinitrobenzene as a cross-linking reagent. J. Biol. Chem. 253:8162–8170.
23. Cuatrecasas, P., and Parikh, I. 1972. Adsorbents for affinity chromatography. Use of N-hydroxysuccinimide esters of agarose. Biochem. 11:2291–2299.
24. Orr, G. A. 1981. The use of the 2-iminobiotin-avidin interaction for the selective retrieval of labeled plasma membrane components. J. Biol. Chem. 256:761–766.
25. Lui, F. T., Finnecker, M., Hamaoka, T., and Katz, D. 1979. New procedures for preparation and isolation of conjugates of proteins and a synthetic copolymer of D-amino acids and immunochemical characterization of such polymer. Biochemistry 18:690–697.
26. Youle, R. J., and Neville, D. M., Jr. 1980. Anti-Thy 1.2 monoclonal antibody linked to ricin is a potent cell-type-specific toxin. Proc. Natl. Acad. Sci. USA 77:5483–5486.
27. Mishell, B. B., and Shiigi, S. M. Selected Methods in Cellular Immunology, Freeman, San Francisco, 1980.

28. Mirro, J. Jr., Schwartz, J. F., and Civin, C. I. 1981. Simultaneous analysis of cell surface antigens and cell morphology using monoclonal antibodies conjugated to fluorescent microspheres. J. Immunol. Method 47:39–48.
29. Khaw, B. A., Scott, J., Fallon, J. T., Cahil, S. L., Haber, E., and Homcy, C. 1982. Myocardial injury: quantitation by cell sorting initiated with antimyosin fluorescent spheres. Science 217:1050–1053.
30. Civin, C. I., and Banquerigo, M. L. 1983. Rapid, efficient cloning of murine hybridoma cells in low gelation temperature agarose. J. Immunol. Meth. 61:1–8.
31. Vienken, J., and Zimmermann, U. 1985. An improved electrofusion technique for production of mouse hybridoma cells. FEBS Lett. 182:278–280.
32. Wojchowski, D. M., and Sytkowski, A. J. 1986. Hybridoma production by simplified avidin-mediated electrofusion. J. Immunol. Method 90:173–197.

INDEX

Note: Italicized page numbers indicate illustrations

A

ACE (angiotensin converting enzyme), 100
Ahkong, Q. F., 28
Alconox (VWR), 51
Aligned cells, and pulse trains, 24
Alignment, 26–27
    observation of in open chamber fusion, *65, 66, 67*
    postfusion, 62
Alignment off time (AOT), 14, 60, 67
Alignment time, variation of in fusion, 48
Amplitude
    electric field strength as, 41
    supracritical, 24
    variation in, 14
Antigen conjugate
    characterization of activity of, *96*
    selective targeting using, 90–93
    testing of, 95
Antihuman chorionic gonadotropin (HCG) antibodies, *94*
Assay
    direct conjugate binding, *95*
    immunofluorescent, *94*
Assymetric breakdown
    factors controlling, 26
    and media composition, 25–26
Avidin, tight binding of with biotin, 90
Avidin-antigen conjugates, construction of, 92–93
Avidinylation, determining extent of, 98

B

B cells
    choice of DNA sequences for targeting, 85
    for fusion, 55–56

immortalization of, 32
    need to activate before electroportation, 77
    recognition of, 67–68
β-HCG (β-human chorionic gonadotrophin), 100
B lymphocytes
    importance of DNA sequences with immortalization functions to electroportation of, 76
    need to activate before electroportation, 84
    transfection of by electroporation, 82–84
Bilayer, resealing of, 24
Biojet CF
    power supply of, 5–6
    use of in electrofusion, 4–6, *5, 15*
Biojet MI, 2, *3*
    features of, 3–4
    injection chamber for, *4*
    power supply of, 3
    use of, 11, 12
Biotin-antigen conjugates, construction of, 93
Biotinylation
    of antigens, 93
    of myeloma cells, 96–98, *97*
BJAB, electroporation with, 40
Breakdown
    assymetric, 25–26
    as rapid, 23
Breakdown pulse, 6
    calculation of field strength of, 18–21
Breakdown time, 24
Breakdown voltage, as function of temperature, 21–22
Brownian motion, 24
Bubbles
    danger of gas, 12
    prevention of air, 58, 60, 65
Buffer, varying ionic strength of, 73

C

Capacitor discharge method, 73
  See also Exponential decay pulse
Cell(s)
  biotinylation of, *97*–98
  for (electro-) fusion, 10
  electrofusion of osmotically prestressed, 15–17
  electroporation of human lymphoid, 31–46
  fibroplastic, and DNA, 43
  for fusion, 55–68
  from immune organs, in vitro activation of, 78
  for injection, 9–10
  mammalian, 32. *See also* Mammalian cells
  preparation of for fusion, 96–99
  preswelling of, and permeabilization, 26
  volume of, and dielectrophoretic forces, 27
Cell chain, formation of, 27
Cell concentrations, effect of on transfection efficiency, 41
Cell condition, as factor in high-efficiency electroporation, 74
Cell count, in microfusion, 63
Cell density
  effect of in open chamber fusion, 65
  in microfusion, 63
Cell fusions, comparative efficiency of PEG and electric field-induced, *49,* 100–101
Cell lines, for gene transfection, 34
Cell passage medium, for electroinjection, 7
Cell ratios, in open chamber fusion, 66–67
Cell recovery, and electric field strength, 81
Cell separation, procedure for, in gene transfection, 34
Cell-to-cell contact, promoting selective, 90
Cell toxicity, as problem, 73
Cell viability
  in biotinylation of cells, 98
  determination of, 79, 80, 81
  and electric field strength, 81
  and electroporation medium, 73
  in electroporation medium/buffer, 41
  and estimating electroporation voltage, 80
Cellular senescence, 76
Chelex beads, as special media, 7
Chinese hamster ovary (CHO) cells, 43
Complete growth medium, 53
  for electrofusion, 8
  for electroinjection, 7
Conical tubes, use of in fusion, 57
Conjugate, testing, 93–96
Cotransfection, efficiency of, 74
Coulter counter, as particle volume analyser, 19
Coulter chamber, use of, 11

Cross-linkers
  heterobifunctional, 92–93
  homobifunctional, 92

D

DC voltage
  effect of on cells being fused, 61
  in microfusions, 63
  in open chamber fusion, 65, 66–67
  variation of in fusion, 48
DEAE-dextran, 32
  neutralization of, 72
Deoxyribonuclease-1 (DNase), 97
DFDNB, 92
Dielectric constants, as functions of frequency of alternating field, 27
Dielectrophoresis, 27, 90
Diffusion, uptake by, 23
DMSO, 32
DNA
  conformations of, 74
  and enhancers, 76–77
  exogenous, as introduced into cells, 72
  and fibroplastic cells, 43
  genomic, purification of, 35–36
  quantitation of, 36
  use of, 11
DNA concentration, effect of on transfection efficiency, 41
DNA conformation, as influence on transfection efficiency, 74
DNA dot blot, 43
DNA electroinjection, 26
DNA molecules, electroinjection of, 7
DNA sequences
  with immortalization functions, 76
  for targeting B cells, 85
DNA transfection
  electroporation-mediated, 32
  transformation of mammalian cells by, 75
DNA transfer, methods of, 43
Dominant acting selectable gene, 75
Dose-response curve, 79, 80
Drug concentration, determination of optimal, 78–79
"Dry" pellets, use of, 58, 62, 64
(Dulbecco's) modified Eagle's medium (DME), 97
Dye(s)
  as probe molecules, 26
  uptake of, and permeabilization, 20

E

EBV-activated B cells, fusions with, 61
  See also Epstein-Barr virus

EBV activation, 56
EDTA, removal of before electroinjection, 7
EIAs, 93
Electric field strength
    as expressed in V/cm, 72–73
Electric field strength, as parameter in
        electroporation, 72
Electric field pulse, potential of techniques of,
        2
Electric field strength and resistance of
        electroporation buffer, 80
    See also field strength
Electrical breakdown, consideration for,
        18–27
    See also Breakdown
Electrodoma, 72
Electrofusion (electric field-induced cell
        fusion)
    of activated human B cells and monse-
        human fusion partners, 48–70
    advantage of technique of, 68
    avidin-biotin mediated, hybridoma
        production by, 89–102
    cause of failure of, 15
    cells for, 10
    equipment for, 6, 50–53
    field strength required for, 20
    in hypo-osmolar solutions, 27
    in iso-osmolar solutions, 28
    media for, 8–9
    of osmotically prestressed cells, 15–17
    vs. PEG fusions, 90, 101
    principles of, 2
    and pulse duration, 23–24
Electroinjection
    assymetric breakdown in, *25*
    cells for, 9–10
    equipment for, 2–4
    field strength required for, 20
    in hypo-osmolar solutions, 27
    media for, 7–8
    miscellaneous equipment for, 6
    principles of, 2
    and pulse duration, 23–24
    temperature for, 23
Electroporation
    as alternative method of generating
        immortalized lymphocytes, 72
    conditions, 39, 42–43, 80
    defined, 38
    efficiencies of, as compared to chemical
        methods, 84
    efficiency of, as compared to micro-
        injection, 84
    electrical and biological parameters for, 72
    generation of immunoglobulin-secreting
        recombinant cells through, 83–84

of human lymphoid cells, 31–46
model system for, 40–42, *41*
procedure for, 39
with targeted liposomes, 39–40, 43
with targeted liposomes, model system for,
        40–*42*
technology for, 43–44
and various cell types, 42–43
Electroporation chamber, requirements of, 73
Electroporation conditions, determining
        optimal, 80
Electroporation voltage, and cell viability, 80
ELISA, 83
Enhancer(s)
    function of, 76, 77, 85
    mechanism of action of, 76
Enzymatic treatment medium, for
        electroinjection, 8
Enzyme immunoassay (EIA), 15
Enzyme-linked immunosorbent assay
        (ELISA), 83
Enzyme-linked immunoabsorbent assay
        (EIAs), 93
Epstein-Barr (EBV),
    infection with, 32, 35
    transformation by, 48
Equator, and generated membrane potential,
        18
Erythrocyte ghosts, 32
Eukaryotic cells, and electroporation, 84
Eukaryotic genes, and mammalian cells, 85
Exponential decay pulse, 73
Exponential decay waveform, characteristics
        of, 73

F

Feeder layers, need for in fusion, 56
Fetal calf serum (FCS), 53
FIA, 83
Field probe particles, use of for
        dielectrophoresis, 27
Field strengths
    determining optimal, Method I, 80–81
    determining optimal, Method II, 81–82
    as function of cell radius, *20*
    and permeability, 18
    supracritical, 19
    See also Electric field strength
"finger flicking," 99
Fluorescence-activated cell sorter (FACS), 97
Fluorescence immunoassay (FIA), 83, 93
Fluorescent microsphere assay, 94–95
Freely suspended cells, and pulse trains, 24
Freunds's adjuvant, 99
Fusion
    cells for, 55–68

Fusion, *cont'd.*
  combining components for, 99
  in helical chambers with L3 fusion
    medium, 61–63
  in helical chambers with Zimmerman
    fusion media, 57–61
  in open chamber, 64–68
  preparation of cells for, 96–99
Fusion chambers, 6, 51–52
  helical, 52
  open, 51
Fusion medium, for electrofusion, 9
Fusion medium (ZFM), 54
Fusion program, steps of, 4, *5*
Fusion ratio, determination of, 60, 61

G

G 8 Hybridoma cells
  for electrofusion, 10
  hypo-osmolar electrofusion of, 14–15
Gene transfer
  by coprecipitation, 72
  by electroporation, 72
  retrovirus-mediated, 84–85
Generated membrane potential
  as angle-dependent, 18–20, *19*
  frequency dependence of, 26
  as radius-dependent, 18, *19, 20*
Geneticin. *See* G418
Gene transfection
  into mammalian cells, 32
  materials/methods, 33–34
G418 (geneticin), 37, 75, 78–79
  titration of, 79
GMP (guanine monophosphate), 37
GMP (guanosine monophosphate), 75
gpt gene, use of in electroporation, 75
gpt selection, 79
  medium for, 84
Guanylic acid, 75

H

HAT media, 90
  hybridoma selection in, 72
HAT/Ouabain 10 M, 54
HAT-sensitivity, verification of, 10
HCMV antigen, plus PWM stimulation, 56
Helical chamber(s) *52*
  fusions in with L3 fusion medium, 61–63
  fusions in with Zimmerman fusion media,
    57–61
  in styrofoam rack, *58*
Helix, washing cells out of, *59*
Hematopoietic cells, and electroporation
    buffer, 74

Heterobifunctinal cross-linker, sulf-hydryl-
    reactive, 92–93
Homobifunctional cross-linker, amino-
    reactive, 92
HT medium, 54
Human B cells, techniques for
    immortalization of, 32
  See also B cells
Human cytomegalovirus (HCMV), 48
Human lymphocytes, generating immortalized
    immunoglobulin-secreting, 71–88
  See also Lymphocytes
Human monoclonal antibodies (HMAbs)
  formation of hybridomas secreting, 47–70
  formation of hybridomas secreting specific,
    68
  generation of, 32
  production of, 48, *49*
  See also Monoclonal antibodies (mABs)
Hybrid
  schedule for feeding of in fusion, 56
  yield of by electrofusion, *16, 17*
Hybridoma(s)
  classical technique of producing, 72
  formation of cells, 2
  formation of secreting human monoclonal
    antibodies, 47–70
  generation of, 82–84
  from hypo-osmolar electrofusion, 13
  yield of by electrofusion, *16*
Hybridoma growth, detection of, 100
Hybridoma production
  by avidin-biotin mediated electrofusion,
    89–102
  by bioselective electrofusion, *91*
  human, 68
Hybridoma technology, and tumor cell fusion
    partner, 32
Hypo-osmolar solutions
  electrofusion in, 13–15
  electroinjection in, 11–13

I

Ig G-type antibodies, 100
Ig secreting hybridomas, 68
Iminobiotin-Sepharose, 92
Immortalizing genes, 76
Immortalizing plasmid, cotransfection of with
    marker plasmid, 82
Immunoglobulin, detection methods for
    production of, 83–84
Immunoglobulin enhancers, 76
Immunology, and recombinant DNA
    technology, 85
Immunotherapy, potential for, 85
IMP dehydrogenase, function of, 75

In vitro activation
    of cells from immune organs, 78
    expansion of B-cell pool by, 48
    as needed for efficient electroporation, 77
    of peripheral blood cells, 77–78
In vitro activated B cells, 55
In vitro immunization, 35
In vivo activation, as needed for efficient
        electroporation, 77
In vivo activated B cells, 55
    for electrofusion, 10
Ion, movement of, 21
Ionic concentration, effect of on time
        constant, 73–74
Iscove's Modified Dulbecco's Medium
        (IMDM), 53
Iso-osmolar electrofusion, of pre-stressed vs.
        untreated cells, 17

K

K6H6/B5 cells, 55
    fusions with, 61
Keyhold lampret hemocyanin (KLH), 99

L

L3 fusion medium, 49, 54–55
    fusions in helical chambers with, 61–63
    microfusions in, 63–64
Laser microbeam, 32
Linearized DNA conformation, 74
Lipids, for preparation of liposomes, 38
Lipopolysaccharides (LPS), 77
Liposomes
    binding of targeted to target cells, 40
    electroporation with targeted, 39–40
    preparation of targeted containing DNA,
        37–38
    use of, 72
Log phase growth, cells in, 9, 77
Lucy, L. A. (or J. A.?), 28
Lymphocyte(s)
    fusion of with mycloma cells, 15, *16*
    importance of stimulation/activation to, 35
    in vitro stimulation/activation of human,
        34–35
    preparation of human, for gene transfection,
        34
    steps in immortalization of 32, *33*

M

Macromolecules, uptake of, 24
Macrophages
    electrotransfection of, *14*
    for injection, 9–10

Major histocompatibility complex (MHC), 37
Mammalian cells
    electrofusion of, 13–15
    electroinjection of, 11–13
    and gene expression, 85
    gene transfection into, 32
    transformation of by DNA transfection, 75
Marker genes, 37
Marker plasmid, cotransfection of with
        immortalizing plasmid, 82
Medium
    for electrofusion, 8–9, 53–55
    for electroinjection, 7–8
    resistance of, and effect on current
        generated, 73
    sets of used in fusion, 49
Medium conductivity, role of in membrane
        charging, 21
Membrane(s), effect of temperature on
        physical properties of, 75
Membrane charging, and role of medium
        conductivity, 21
Membrane contact, methods for promoting,
        90
Mice, immunization of, 99
Micro Helical Chamber, 63
Microfarads, 73
Microfusions, 63–64
Microinjection
    direct, 32
    of DNA, 72
    efficiency of, as compared to
        electroporation, 84
Microspheres, 94–95
Mitosis, and DNA uptake, 74
M-maleimidobenzoyl N-hydroxy-succinimide
        (MBS), 92
Monoclonal antibodies (mABs), 37
    electro vs. PEG fusion method of
        production of, 101
    generation of human, 32
    production of, *91*
Mouse-human fusion partners, and
        hybridomas secreting human
        monoclonal antibodies, 47–70
Mouse-human heteromyeloma
    as fusion partners, 55
    as hybrids, 67
    parameters for fusion with, 51
    use of, 48, *49*
Mouse L cells
    electroinjection of, *12*
    electrotransfection of, *13*
    for injection, 9
Mouse myeloma cells, for electrofusion, 10
Murine monoclonal antibodies, production of,
        48

Murine transformants, formation of, 2
Myc expression, effects of, 76–77
Myc oncogene, coupling of with enhancer, 85
Mycophenolic acid
    as medium, 37
    as metabolic inhibitor, 75–76
    titration of, 79–80
Myeloma cells
    antigen coated, 90
    avidinylation of, 98
    biotinylation of, 96–98, 97

N

NEO marker gene, 37
    use of in electroporation, 75
Neo gene selection, 78
    medium for, 84
Neomycin, and mammalian cells, 75
N-hydroxysuccinimidyl (NHS), 96

O

Oncogene technology, and immortalization of
    human B cells, 32
Onogenic DNA/oncogenes need for source of,
    43–44
Open chamber, advantage of, 64
Open chamber fusion, appraisal of, 67–68
Osmolarity, effect of on time constant, 73–74
Osmotic shocks, 17

P

P 3 (P3X63/Ag 8.653) cell line, 96
Pelleting, of cells in fusion, 58
Peripheral blood cells
    activation of, 77
    in vitro activation of, 77–78
Permeability, increase in, as function of field
    strengths, 18, 20
Permeabilization
    angular dependence of, 20
    and preswelling of cells, 26
Phenol red, toxicity of, 7
Phosphate buffered saline (PBS), 96
Phytohemaglutinin (PHA), 77
Planar membranes, and generated membrane
    potential, 18
Plant cells, and study of spatial
    permeabilization, 28
Plasmalemma membrane, breakdown voltage
    of, 26
Plasmid(s)
    and calculating transfection efficiency, 81
    electroinjection of, in hypo-osmolar
        solutions, Method I, 11–13

linearization of, 11
simultaneous transfection of separate, 74
types of, 81, 82
use of in hybridoma generation, 82
use of linearized for stable transfection, 81
Plating
    in fusion, 57
    volume of HT needed for, 60–61
Plating density, determination of, 60, 61
Pokeweed mitogen (PWM), 48
Poles, and generated membrane potential, 18
Polyethylene glycol (PEG)
    as fusogen, 48, 49
    procedure, 72, 90, 101
Pore closure, 42–43
Post-fusion medium (ZPFM), 54
    for electrofusion, 9
Post-pulse medium, for electroinjection, 8
Potential across the membrane, constituents
    of, 25
Prokaryotic genes, expression of, 85
Proteins, disruption of by osmotic means, 27
Protoplast fusion, use of, 72
Pulse(s)
    application of for transfection, 11
    number of as parameter, 41
    types of, 72–73
Pulse duration, 23–24
Pulse medium, for electroinjection, 8
Pulse sequence, importance of correct, 24
Pulse trains, 24–25
Pulse width, duration of pulse as, 41
Pulsing, and gas bubbles, 12
Purine nucleotide synthesis, inhibition of, 76
Purine salvage pathway, 76, 80
PWM stimulation, 56

R

Radioimmunoassay (RIA), 83, 93, 95
Recombinant cells, generation of
    immunoglobulin-secreting through
    electroporation, 83–84
Recombinant DNA technology
    and DNA sequences, 85
    generating immortalized immunoglobulin-
        secreting human lymphocytes by,
        71–88
Relaxation time
    defined, 21
    as function of cell radius, 21, 22
    as function of external conductivity, 21, 22
Resealing
    as function of temperature, 22–23
    importance of, 24
Resealing medium, for electroinjection, 8

Resistance
    within electroporation chamber, 73
    of medium, effect of on current generated,
        73
    and varying salt concentrations, 74
Resistance-capacitance (RC) time constant,
    73
    See also Time constant
Resistance measurements, functions of, 6
Retrovirus-mediated gene transfer, efficiency
    of, 84–85
Room temperature
    as rule for electroporation of mammalian
        cells, 75
    as rule to increase transfection efficiency,
        84. *See also* Temperature

S

Salt, varying concentrations of and resistance,
    74
SBC-H20 cells, 55
Selection medium
    for electrofusion, 9
    for electroinjection, 8
Sendai virus envelopes, reconstituted, 32
Shearing forces, sensitivity of permeabilized
    cells to, 23
Smith, Kline French laboratories, 32
SPDP (N-succinimidyl 3-(2-pyridylthis)
    propionate), 98
Spectrophotometer curvettes, 73
Spherical cells
    calculation of field strength of breakdown
        pulse for, 18–21
    and generated membrane potential, 18
Spleen cells
    in vivo priming of, 98
    isolation of, 99
Square wave pulse, 73
Stationary state, meaning of, 20–21
Streptomyces avidinii, 93
Supercoiled DNA conformation, 74

T

Target cells, binding of targeted liposomes to,
    40
Targeting, selective, using antigen conjugates,
    90–93
Temperature, 21–23
    control of in electroinjection, 21–22
    effect of on duration of permeabilized
        stated, 75
    effect of on electrofusion, 90
    effect of on membranes, 75

perturbations as function of, 22–23
    See also Room temperature
Time constant, in electroporation, 73
Transcriptional activators, 76
Transfectants, monoclonal antibody
    production/human, 43–44
Transfection
    artificial, 32
    DNA, 32
    gene,32
    natural, 32
    protocols for, 38–40
Transfection efficiency
    determining maximum, 82
    and electroporation medium, 73
    factors affecting, 85
    and room temperature, 84
Transfer
    facilitator-aided, 32
    vehicle-mediated, 32
Transformants, vectors and selection for, 37
Transmembrane pores, formation of, *19*
Tumor cell, as fusion partner, 32, 35

U

University of Würzburg, 2
Uptake, and viability, 23

V

Viral vectors, use of, 72
Voltage, variation of in fusion, 48
    See also DC voltage

W

Wash medium, for electroinjection, 8

X

XGPRT (xanthine-guanine
    phosphoribosyltransferase)
    as gpt gene, 75
    as marker gene, 37, 40, 43
XMP (xanthine monophosphate), 37
XMP (xanthylic acid), 75
XPRT as gpt gene,75

Z

Zimmermann Cell Fusion System, 48, 50–51
Zimmermann fusion medium (ZFM), 49, 54
Z 1000 Zimmerman Cell Fusion System, 50